COSMIC
PARADOXES

COSMIC PARADOXES

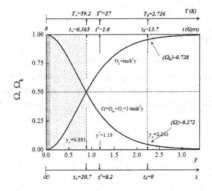

Julio A Gonzalo

Universidad Autonama de Madrid, Spain

World Scientific

NEW JERSEY · LONDON · SINGAPORE · BEIJING · SHANGHAI · HONG KONG · TAIPEI · CHENNAI

Published by

World Scientific Publishing Co. Pte. Ltd.
5 Toh Tuck Link, Singapore 596224
USA office: 27 Warren Street, Suite 401-402, Hackensack, NJ 07601
UK office: 57 Shelton Street, Covent Garden, London WC2H 9HE

British Library Cataloguing-in-Publication Data
A catalogue record for this book is available from the British Library.

ISBN-13 978-981-4355-61-2
ISBN-10 981-4355-61-5
ISBN-13 978-981-4355-11-7 (pbk)
ISBN-10 981-4355-11-9 (pbk)

Typeset by Stallion Press
Email: enquiries@stallionpress.com

Printed in Singapore.

To my parents, and to Blessed John Paul II

Foreword

Paradoxes — either real or apparent contradictions — have been part and parcel of cosmology for centuries. Many have tried to interpret the dark night paradox (Olber's paradox) in such a way as to save the presumed infinity of the universe with no success.

In this book cosmic paradoxes are seriously examined from an honest, conservative point of view. What does a conservative approach mean? It is an approach which does not impose *a priori* conditions (such as the necessary flatness of space-time all the way through cosmic expansion) disregarding the evidence to the contrary if there is any. And it is an approach which, on confronting any objective entity, ranging from an atom to a galaxy and even the whole cosmos, does not take lightly *energy conservation*. This was done in the 1950s by Gold, Bondi and Hoyle, in their *Steady State Theory*, and in the 1980's, up to now, by Guth, Linde *et al.*, in the *Inflationary Theory*. Of course, if the requisite of energy conservation is relaxed one may enjoy unlimited freedom. But this freedom remains suspicious even after receiving support (real or imaginary) from Quantum Mechanics.

The Big Bang model, anticipated by Georges Lemaitre and formulated some years later by Gamow, Alpher and Herman, is certainly one of the most spectacular scientific successes in the entire history of physics. It remains today, in my opinion, surrounded by too much speculation, without real physical support, for the time being.

Recently, at a scientific meeting in Philadelphia, one attending colleague dared to suggest that the TOE should be called the "Theory *of* Everything *except* Experimental Reality". As Planck once said: "Theorists are numerous and paper is patient". Because of the finiteness of the speed of light, here and now, we are seeing a *superposition* of snapshots, coming from a multitude of galaxies when they were at distances ranging from the relatively close Andromeda to the most distant protogalaxies, or quasars, emitted at times ranging from a few million light years to around ten billion light years. It is therefore absolutely mandatory to take this into consideration if we are to interpret correctly cosmological data.

Stanley Jaki in *The Road of Science and the Ways to God* recalled what Planck took as the paradox of relativity: instead of relativizing everything, it unfolded *absolute*, objective aspects of the physical world. In October 26–27, 1990, a Symposium on *Physics and Religion in Perspective* was held in Madrid, sponsored by BBV and UNED. The opening lecture, by Prof. Stanley L. Jaki (Princeton) was on "Physics and the Universe: From the Sumerians to the late twentieth century". Professor Jaki's lecture is reproduced in an appendix at the end of this book.

In August 16–20, 1993 a Summer Course on *Astrophysical Cosmology*, organized by the Universidad Complutense, Madrid, and co-sponsored by the Spanish Royal Academy of Sciences (Real Academia Española de Ciencias Exactas Físicas y Naturales), was held at the historic site of El Escorial, near Madrid. The Program of the course included talks by a few true luminaries of modern scientific cosmology: Ralph A. Alpher, George F. Smoot, John C. Mather, who acted as Co-Director of the Summer Course, aided by Jose Mª Torroja (Secretary of the Spanish Royal Academy of Science), other distinguished foreign speakers (Jerome Lejeune, Hans Elsaesser, Stanley L. Jaki) and a number of Spanish speakers from different institutions (Jose L. Sanchez Gomez, Antonio Fernandez Rañada, Manuel Catalan and Julio A. Gonzalo). The Summer Course was coordinated by Rodolfo Nuñez de las Cuevas, Ignacio Cantarell and Julio A. Gonzalo, as Secretary, and was whole heartedly supported from the beginning by Miguel Angel Alario, Dean of the Faculty of Chemistry (UCM) and General Scientific Coordinator of the El Escorial Courses that year.

The first draft of the Summer Course Program was sketched in June 1992, on the occasion of a conference on "Is there such a thing as a last word in Physics?", given by Stanley L. Jaki at the Spanish Royal Society, followed by an animated question and answer period. The confirmation of the Planck characteristic distribution of the cosmic background radiation spectrum and the detection of their minute anisotropies measured by NASA's COBE satellite completed in a satisfactory way the description of the initial moments of the universe's thermal history. Some of the protagonists of this story were there, at the El Escorial Summer Course. I think it is fitting to reproduce here in full the English version of the 1993 Summer Course at El Escorial.

16 Monday	10:00	R. A. Alpher (Schenectady, N. Y.): *"The Cosmic Background Radiation (CBR)"*
	11:30	G. F. Smoot (Berkeley): *"Satellite observations of the CBR anisotropy"*
	16:00	Round Table: *"The significance of the CBR"* Moderator: J. M. Torroja; R. A. Alpher, G. F. Smoot, J. C. Mather, H. Elsaesser.
17 Tuesday	10:00	J. L. Sánchez Gómez (Madrid) *"Cosmology and elementary particles"*
	11:30	A. Fernandez Rañada (Madrid) *"Thermal history of the universe"*
	16:00	Round Table: *"Evaluation of the big-bang model"* Moderator: A. Tiemblo; J. L. Sanchez Gómez, A. Fernandez Rañada, J. M. Quintana, J. A. Gonzalo.
18 Wednesday	10:00	J. C. Mather (NASA): *"The COBE project: achievements in perspective"*
	11:30	J. Lejeune (Paris): *"Neurophysiology of human intelligence"*

	16:00	Round Table: *"Life and intelligence in the universe"*
		Moderator: M. Artigas; J. C. Mather, J. Lejeune, S. L. Jaki, T. Vives.
19 Thursday	10:00	H. Elsaesser (Heidelberg) *"The Via Lactea"*
	11:30	M. Catalan (San Fernando, Cadiz) *"The Planet Earth"*
	16:00	Round Table: *"Earth, Via Lactea and Cosmos"*
		Moderator: B. Rodriguez Salinas; H. Elsaesser, M. Catalan, (R. Nuñez de las Cuevas)
20 Friday	10:00	J. A. Gonzalo (Madrid) *"From the Solar System to the confines of the observable universe"*
	11:30	S. L. Jaki (Princeton) *"Cosmology: An empirical science?"*
	16:00	Round Table: *"Physics, Philosophy and the search for meaning in the universe"*
		Moderator: F. Garcia Moliner; J. A. Gonzalo, S. L. Jaki, I. Cantarell, R. A. Alpher.

Brief Summaries of the lectures given by the invited speakers at the Summer Course are given below.

R. A. Alpher: "The Cosmic Background Radiation"

It is commonly accepted that the observable universe evolved from a high density and high temperature primitive state some fifteen to twenty billion years ago. The event ended up being called the "Big Bang". The most successful relativistic model for this universe is based upon the homogeneous and isotropic expansion of an ideal fluid from an initial state dominated by radiation to the present state dominated by matter. The model encompasses the thermonuclear synthesis of light elements during the first few minutes to the transition from an expansion controlled by radiation to other controlled by matter after a few hundred thousand years. In subsequent epochs the present structure of the universe developed and the heavier elements were formed in the stars.

G. F. Smoot: "Satellite observations of the CBR anisotropy"

The Cosmic Background Explorer (COBE) satellite was developed to measure the diffuse infrared and microwave radiation from the early universe, to the limits set by the astrophysical foregrounds. COBE has three instruments: The Differential Microwave Radiometer (DMR) maps the cosmic radiation precisely. The Far Infrared Spectrophotometer (FIRAS) compares the cosmic background radiation spectrum with that of a precise black-body. The Diffuse Infrared Background Experiment (DIRBE) searches the accumulated light of primeval galaxies and stars. FIRAS has measured the cosmic background radiation spectrum nearly 1000 times more precisely than previous observations. The DMR discovered the cosmic background to be anisotropic at a level of one part in 10^5. The results strongly constrain cosmological models and the formation of structure in the present universe. The results from COBE are in good quantitative agreement with the calculations of the hot Big Bang model. The DMR data are in agreement with inflationary model predictions.

J. L. Sanchez Gomez: "Cosmology and elementary particles."

Brief summary of the standard cosmology model: expansion of the universe, cosmic background radiation and primordial nucleosynthesis. Cosmological consequences of particle physics: matter-antimatter symmetry, the problem of the exotic dark matter and the curvature of the universe, phase transitions in the primordial universe and inflationary Big Bang model.

The aim of the talk is to present the last advances in elementary particle physics of interest for cosmology.

The talk will have a markedly informative and descriptive character. Excessively formal questions will be avoided and the fundamental aspects of the intriguing relationship between elementary particles and cosmology will be given, avoiding, as much as possible, technical language.

A. Fernandez Rañada: "Thermal history of the universe"

1. The universe's expansion and Hubble's law: A gas of galaxies; the recession of galaxies; the age of the universe.
2. The explosion: Thermal history of the universe. From $t = 10^{-11}$ s to today; from $t = 10^{-43}$ s (Planck's time) to $t = 10^{-11}$ s; from $t = 0$ to $t = 10^{-43}$ s.
3. Test and problems of the standard Big Bang model. Problems: Singularity; flatness; horizon; galaxy formation.
4. Inflation: Lack of a quantum theory of gravity. How "inflation" trys to solve the above cosmological problems. An emergent theory: The multiverses.

J. C. Mather: "The COBE project: achievements in perspective"

To begin with I would like to point out that we can look back in time by looking at things that are very far away. We simply use the fact that light comes to us at a definite speed and so when we look at the sun, we see the sun as it was 500 seconds before we look at it; when we look at something in the middle of the galaxy we see it as it was 25 000 years ago and when we look at something extremely far away we can see it as it was 15 billion (10^9) years ago.

> *General considerations: No observable center / Hubble's law / Friedmann universes / Dark matter / Galactic rotation / Multiple images of distant galaxies...*
> *Main events: Nucleosynthesis / Number of photons fixed / Decoupling / Galaxy formation...*
> *The three COBE instruments: Differential Microwave Radiometer (DMR) / Far Infrared Absolute Spectrometer (FIRAS) / Differential Infrared Background Experiment (DIRBE)... COBE's scientific team...*

To conclude I would like to talk a little bit about what comes after COBE. We are going to turn off the COBE satellite this December because we got the data we wanted to get. We still have experiments

to look for the anisotropy at the most interesting scale currently, 1°
or 1/2° in size. People are doing balloon experiments going to the
Antarctica, making interferometers to use on the ground, and even
thinking about sending additional space probes out to get away from
the interference of the Earth's atmosphere.

J. Lejeune: "Neurophysiology of human intelligence"

Were the human intelligence the fruit of pure chance, how could it
be able to decipher, however partially, the rules of the universe?

A connivence between the physical/chemical laws and the pecu-
liarities of the substratum of the thought is worth investigating.

The inborn circuitry of our neurons could be our teacher in
elaborating theories. From Euclid's ideal plane to Cartesian coordinates, from Galileo's falling marble to Newton's attraction, from
Maxwell's little devil to Einstein's space-time, the milestones of science have their preexisting anatomical counterparts, macroscopic or
microscopic.

For example, the succession of the mathematical discoveries
occurred in the same sequence as the steps we use in analyzing optical information from the outside world down to the obscure center
that sees.

Discovery or reminiscence, the world of the ideas is not in dishar-
mony with biological observation.

The great caution remains: reason's networks are only one of the
tools at our disposal. The heart, with its built-in memory of life,
should never be overlooked. Separating them (heart and reason) is a
deadly blow to intelligence itself.

H. Elsaesser: "The Via Lactea"

The investigation of the *Via Lactea*, our galaxy, is a classic subject of
study in astronomy with a long history. However, new and interesting aspects come out again and again in such a way that several of
the problems of maximal actuality today, such as the active galactic
nuclei or the search for dark matter, are subjects of equally active
research on the *Via Lactea*, as older problems.

A summary in a few words of what astronomy has learned from research on our galaxy could be the following:

– What are the astral bodies (their physical nature, their formation, their development...)?
– What is a galaxy (its star content, its dimension, its structure...)?

Today there can be no doubt that the system which makes up the *Via Lactea* is one of the large disk-shaped spiral galaxies similar to Andromeda and to a multitude of other like systems which we observe in the universe.

> *Local Group / Generations of stars / Galactic structure / The Hertzsprung–Rusell diagram / Star age at the time of separation from main sequence / M17 and M13 / Distribution of young objects along the Via Lactea spiral / IR measurement on the Via Lactea with IRAS / Spiral structure / Distribution of giant gaseous regions / Orbit of Spherical groups M15 / Velocity rotation as a function of distance to galactic center...*

What type of matter constitutes dark matter? It is not clear. Maybe a multitude of white dwarfs and red stars with brightness too low to be detected by the methods presently used. Another hypothesis is neutrinos with mass or even unknown elementary particles.

M. Catalan: "The Planet Earth"

During the last few decades joint developments in radiostronomy, astrophysics and astronomy have given rise to a complete practical reconsideration of what we know about the universe.

Similarly, the models, on which our knowledge of the formation, interaction and physical structure of our planet were based, have undergone not only profound changes but a veritable revolution of the theories, and concepts that up to now are generally accepted have been substituted by opposite concepts and theories.

Then, for the first time, we can approach a physical picture of the evolutive dynamics of the Earth as a part of a universe in continuous evolution.

In this lecture, the formation of the Earth within the interior of the stellar system is described; the process of formation and dynamic evolution is examined; the special techniques available from the Earth and from outer space using today's astrometry can be used to observe the geometry of the universe through the deformation of extragalactic reference systems.

J. A. Gonzalo: "From the solar system to the confines of the observable universe"

Human ingenuity was able, in ancient Greece, of going from an accurate estimate of the size of the earth to rough estimates of sizes and distances to moon and sun. In medieval times, at the University of Paris, the concept of inertial motion and the realization of the possible rotation of the Earth prepared the way for Newton and for the idea of a cosmic gravitational force.

In the twentieth century, Einstein's Theory of Relativity, the general recession of galaxies, the cosmic relative abundance of light elements and the discovery of the cosmic background radiation paved the way for the big bang concept, earlier envisioned by Lemaitre. Today we can get fair estimates of the scales of masses, distances, and times in the universe, confirmed by observation.

We can conclude that the universe is admirably well done and that human intelligence (not without strenuous effort sometimes accompanied by good luck) is admirably well suited to investigate it. Good astrophysics does not lead to God directly, but it certainly is a useful pointer to the Creator.

S. L. Jaki: "Cosmology: An empirical science?"

Cosmology has become a branch of physical science as avidly cultivated as, for instance, are electromagnetics and atomic physics. This is in a marked difference from the status of cosmology a hundred, let alone two hundred years ago. Still a basic and hardly ever noted difference remains between scientific cosmology and other branches of physics. Unlike these branches, which can empirically demonstrate the existence of their subject matter, scientific cosmology has not yet done so. Its cultivators still have to ponder in depth some such questions: Is the universe the largest sum of galaxies that can actually be observed? Is the universe that sum of material entities whose existence can be inferred on a theoretical basis? Or is the universe an absolute totality of matter? Can such an entity be ever the object of a scientific cosmology which is unambiguously empirical? If not, what are the considerations, be they non-empirical and non-mathematical (and in that sense non-scientific), that can assure one about the existence of such a totality?

In 1993, at the time of the El Escorial conference, the age of the universe was located somewhere between ten and twenty billion years. And the numerical value of Hubble's parameter was estimated between 50 and 100 km/s/Mpc. The interaction with Ralph Alpher, John Mather, Stanley Jaki and others at the conference (and after) was extremely useful in order to make later accurate anticipations of $t_0 = 13.7 \cdot 10^9$ yrs and $H_0 = 65$ km/s/MPC. These accurate estimates were published in 1998 at *Acta Cosmologica (Krakow)*, with the help of my dear Polish friend Dr. Janusz Przeslawski, and some years later reprinted as an Appendix in *Inflationary Cosmology Revisited* (World Scientific, Singapore, 2005).

My colleagues N. Cereceda, G. Lifante and myself did not need any of the artificial assumptions of Inflationary Theory to arrive to those accurate numerical estimates, beautifully confirmed in 2003 by WMAP's data, the NASA satellite successor of COBE.

I would like to thank Manuel M. Carreira SJ Gines Lifante, and Manuel Marques for their encouragement and competent criticism. To Stanley Jaki and Ralph Alpher (both recently deceased) as well as to John C. Mather, Dermott Mullan and Anthony Hewish for their helpful comments and suggestions, and to my former student and collaborator Feli Martínez Ruiz for typing and reviewing the manuscript.

I am grateful also to Prof. K. K. Phua for his editorial advice and generous support.

Julio A. Gonzalo
Madrid, May 31, 2010

Contents

Foreword   vii

List of Tables   xix

List of Figures   xxi

Part I: Facts and Principles   1

1.   Energy Conservation   3
2.   Energy Non-Conservation Means Too Much Freedom   9
3.   The Four Interactions   15
4.   Matter, Radiation, Particles   21
5.   The Dark Night Sky and Olbers' Paradox   27

Part II: Relativistic Cosmology   35

6.   General Relativity and Cosmology   37
7.   The Friedmann–Lemaitre Solutions   45
8.   The Role of Radiation Pressure   53
9.   The Einstein–Lemaitre Correspondence   59

Part III: More Paradoxes   63

10.   The Accelerating Universe Paradox   65

11. The Missing Mass and Dark Energy Paradoxes   71

12. The Photon-to-Baryon Ratio Paradox   77

Part IV: A Contingent Universe   **83**

13. The Universe is Finite, Open and Contingent   85

14. The Very Early Universe: Indeterminacy
     or Uncertainty   89

15. Singular Moments in Cosmic History   93

16. A Brief Outline: World Events and Cosmological
     Discoveries from -4500 to 2010   99

Appendix   103

Glossary   121

Author Index   127

Subject Index   129

List of Tables

Table 3.1.   Cosmic quantities in Planck's units 18

Table 4.1.   Subatomic particles. 23

Table 11.1. The evolution of Ω_m (matter), Ω_r (radiation) and Ω_k (space-time curvature potential energy) as a function of time t (Gyrs) and temperature $T(\mathrm{K})$. 74

Table 13.1. Compton and Schwarzschild radii for massive objects 86

List of Figures

Fig. 6.1   Deviation of light ray from a star when it passes near the Sun's surface: $\alpha \approx sGM/Rc^2$ where s is a dimensionless factor of order unity, G Newton's constant, M the Sun's mass, R the Sun's radius and c the velocity of light.   40

Fig. 7.1   Contributions to the density parameter Ω_m (matter), Ω_r (radiation), Ω_k (space-time curvature) as a function of y (dimensionless cosmic parameter), z (redshift), t (time) and T (CBR temperature).   51

Fig. 8.1   Planck's black body radiation spectrum giving the relative (monochromatic) brightness in ergs/cm^2·sec as a function of the wavelength in cm. The CBR observed by Penzias and Wilson in 1965 at 7.3 cm corresponded to a characteristic temperature of about 3.5 K.   54

Fig. 10.1 Apparent recession velocity, $\dot{r}/c = v/c$, versus apparent distance r(pc), fitted by $v/c = (v_m/c)$ $\tanh(r/r_m) = (v_m/c)\tanh(\gamma r)$, where $\gamma \equiv 1/r_m$. Keeping only the nine points pertaining to the "High z Supernova Search" results in a much better fit with the above equation; this is not shown in Fig. 10.1 because α in Eq. (10.6) is somewhat larger than α in Eq. (10.7).   69

Part I: Facts and Principles

1. Energy Conservation
2. Energy Non-Conservation Means Too Much Freedom
3. The Four Interactions
4. Matter, Radiation, Particles
5. The Dark Night Sky and Olbers' Paradox

As a firm starting point we are left with only a few principles, among them the universal law of energy

Max Planck
(1858–1947)

Antrittsrede zur Aufnahme in die Academie vom 28 Juni 1894, Physikalische Abhandlungen, 3:4, p. 3

Chapter 1

Energy Conservation

One of the most dramatic achievements of nineteenth century physics[1] was the enunciation of the principle of *energy conservation*. Clearly, this fundamental principle has very profound consequences in every branch of physical science.

Nineteenth century physicists realized that burning coal could set a steam machine in motion; and that this motion was not in any sense a creation out of nothing, but the result of a direct transformation of energy from one form into another. Therefore, mechanical work could be done only because a certain amount of energy was previously stored into the burning material, but only up to a certain amount, and no more.

Further, the work of a steam-engine itself could be translated into electrical energy which, then translates into heat or light, both of which, again, into chemical or mechanical energy. The total amount of energy could not be increased or diminished, in any way whatsoever, in any of these transformations. This is the basic meaning of the principle of energy conservation.

Haeckel,[2] a prominent German nineteenth century materialist, dared to combine the principles of matter conservation and energy conservation to set forth what he called the fundamental principle of the Monist Religion of the future.

3

He meant by it that this fundamental principle was in direct contradiction with the Christian views, which saw man and cosmos as created by an intelligent and all-powerful Creator. His was, and still is, a gratuitous assertion: What doctrine of the Church did forbid accepting the principle of matter and energy conservation, if the evidence in its favor should seem conclusive? Should Christianity be reproached for not teaching this principle in the "Credo"?

Of course, Christian Revelation was not set forth to teach man physics or chemistry, or geology, or biology as such, i.e., to teach truths of the natural order. As no less an authority than Saint Augustine said about sixteen centuries ago, Revelation tells us, "not how heavens go, but how to go to heaven".

The whole four thousand years old Judeo-Christian tradition, and especially the two thousand years old Catholic tradition, is witness of a staunch defense of reason as a God given gift to man, so that man may be able of acquiring truths of the natural order. But reason should not be confused with revelation, which certainly can go beyond reason, but never directly against it to the extent of contradicting it flatly. Reason can help revelation. In fact it is clearly needed to properly understand Revelation. And Revelation, if not abused, and properly respected, may help to get insights into the natural order insights which otherwise might be missed.

According[3] to Monsignor Georges Lemaitre (1894–1966), the father of the Big Bang model:

> *Both — the believer and the non-believing scientist — will tray to decipher the complex palimpsest of nature in which traces of the long evolution of the world are confused and mixed up with one another. The believer may have the advantage of knowing that the enigma has a solution, that the written text underlying it is, after all, the work of an intelligent being, given that the problem proposed by nature has been posed to be solved, and that, no doubt, its difficulty is proportioned to the capacity of men now or in the future. This will not give him, perhaps, more resources in his research, but it will contribute to maintain him in that healthy optimism without which no sustained effort will last for very long.*

Is it too much to ask from contemporary cosmologists full respect for the principle of energy conservation?

Sir Arthur S. Eddington, mentor of Georges Lemaitre and himself one of the greatest contributors to early twentieth century cosmology said:

> *We are of course allowed to rearrange the matter of the universe ... but in such rearrangement the experimenter cannot, and the theorist must not, violate the conservation of energy.*

The universe we see today, through ground-based powerful telescopes and through space telescopes (like NASA's Hubble telescope), is made up of a spherical distribution of galaxies (truly enormous but finite), of the order of 10^{12}, receding radially from one another at very large speeds within an expanding, homogeneous and quasi-isotropic spherical cloud of radiation, both centered at a certain point in space-time, a point which could be properly called the center of mass of this finite universe.

13.7 Gigayears ago,[4] more or less, all cosmic matter and all cosmic radiation were concentrated at this singular point in space-time, the center of mass of the universe. The Big Bang must have taken place precisely then and there. Matter and energy may reasonably be assumed to be conserved thereafter up to present times and beyond. As will be seen below, the uncertainty principle does not allow us to say anything before a time of the order of 10^{-103} sec after the Big Bang.

Einstein's original[5,6] cosmological equation for a galaxy of mass M_G receding from the center at maximum speed, $\dot{R}$, located therefore at maximum distance of R from it, can be written as

$$\frac{1}{2}M_G\dot{R}^2 = M_G\frac{4\pi}{3}G\rho R^2 - \frac{1}{2}M_G k c^2 + \frac{1}{2}M_G\frac{\Lambda}{3}R^2 c^2.$$

The term on the left-hand side is the kinetic energy and the three right-hand side terms are, respectively, the *gravitational potential energy*, the (relativistic) *space-time potential energy*, and the so-called *cosmological constant potential energy*, originally introduced by Einstein to counter gravitation and to produce (what he then through was) a "static" universe. Because his previous success in applying general relativity to solve problems, such as the bending of stellar light passing close to the Sun's surface, were *all* cases in which the space-time curvature was positive ($k > 0$), he was expecting $k > 0$

also in his cosmological equation, and he needed a cosmological constant term to counter gravity and produce a static universe. It was realized later (after the discovery of the general galactic recession) that the second term with $k < 0$ instead of $k > 0$ could play basically the same role as a third term with $\Lambda > 0$ in Eq. (1.1). Einstein, then, disavowed his cosmological constant. In his letter to Lemaitre[7] dated September 26, 1947, he said:

> *Since I have introduced the term (the "cosmological constant" term)*
> *I had always a bad conscience. But at the time I could see no other*
> *possibility to deal with the fact of the existence of a finite mean*
> *density of matter. I found it very ugly indeed that the field law of*
> *gravitation should be composed of two logically independent terms*
> *which are connected by addition ... I am unable to believe that such*
> *an ugly thing should be realized in nature.*

Lemaitre, on his part, did not agree with Einstein, and saw in this "cosmological constant" term perhaps a reasonable escape to a serious problem he then had: trying to explain a too-short cosmological age of the universe based upon the current Hubble's wrong estimate of the galactic distance scale. The problem was definitely solved half a century later when better determinations of the scale of galactic distance made compatible dynamic cosmological estimates of the age of the universe, and with the ages of the oldest globular star clusters known (turning off main sequence).

As it will be shown in greater detail in Part II, the *k-term* (space-time curvature term) and the Λ-*term* (cosmological constant) can be shown to play basically the same role in the solutions of Einstein's cosmological equations. Therefore, it is adequate to proceed, as did Friedmann, setting $k < 0$ and $\Lambda = 0$, or equivalently, setting $k = 0, \Lambda > 0$, in order to obtain the dynamic evolution of an open universe. However, as also pointed out in Part II, $k < 0$ can be given a straightforward physical meaning linked to the radiation pressure associated to the cosmic background radiation, which is not so easy with $\Lambda = 0$.

It will be seen that dividing both sides of Eq. (1.1) by $\frac{1}{2}M_G\dot{R}^2$ we get

$$1 = [\Omega_m(t) + \Omega_r(t)] + \Omega_k(t) + \Omega_\Lambda(t). \qquad (1.1)$$

Here $\Omega_m(t) = \rho_m(t)/\rho_c(t)$ is the dimensionless *matter* mass density parameter, obtained dividing the actual (time-dependent) matter mass density by the so-called critical (also time-dependent) matter mass density, corresponding to $k = 0$; $\Omega_r(t)$ is the radiation mass density parameter obtained likewise, which is at present much lower, but it was equal to it at decoupling (matter/radiation equality, atom formation time); $\Omega_k(t)$ is the time-dependent space-time *curvature* density parameter, and $\Omega_\Lambda(t)$ is the cosmological constant density parameter, which can be ignored for the moment for practical purposes, as did Friedmann. We can set $\Omega = \Omega_m + \Omega_r$ for convenience.

It will be seen that, using the compact Friedmann–Lemaitre solutions for an open $k < 0, \Lambda = 0$ universe, we get

$$t = \frac{R_+}{c|k|^{1/2}}[\sinh y \cdot \cosh y - y] \tag{1.2}$$

$$R = R_+ \sinh^2 y \tag{1.3}$$

with $2G(\frac{4\pi}{3}R^3\rho) = \text{constant} = c^2|k|R_+$. These solutions lead to the value of $\Omega(t)$ coming down from $\Omega(0) = 1$ (at very early times) to $\Omega(t_o) \approx 0.044$ (at present), and for $\Omega_k(t)$, coming up from $\Omega_k(0) = 0$ to $\Omega_k(t_o) \approx 0.965$ at present.

It will be seen below that this is relevant for finding a reasonable explanation for the large missing dark mass and the large missing dark energy problems.

Energy conservation from the very beginning to the present is of course compatible with a change in the equation of state of the universe at the transition from a *plasma universe* made up of nuclei (mainly H and ^{4}He) and electrons, to an *atomic universe* made up of atoms, then made up of cosmic dust, and then of stars and galaxies, as discussed in Part III.

References

1. K. A. Kneller, *Christianity and the Leaders of Modern Science*, with an introductory essay by Stanley L. Jaki (Real View Books, Fraser, Michigan, 1995), p. 7.
2. E. Haeckel, *Die Zukunft III.* (Berlin, 1895), p. 199.

3. O. Godart-M. Heller, *Les Relations Entre la Science et la Foi Chez Georges Lemaitre*, in Pontificia Academia Scientiarum Commentarii, Vol. III, No. 21, p. 11.

4. J. A. Gonzalo, *Inflationary Cosmology Revisited* (World Scientific, Singapore, 2005), Appendix, p. 83.

5. S. Weinberg, *Gravitation and Cosmology*, (Wiley and Sons, New York, 1972).

6. S. Weinberg, *Cosmology* (Oxford University Press, Oxford, 2008).

7. J. Farrell, *The Day Without Yesterday* (Thunder's Mouth Press, New York, 2005), p. 169.

Chapter 2

Energy Non-Conservation Means Too Much Freedom

We may begin by noting that the principle of energy conservation has not been consistently honored by some prominent physicist from very early in the twentieth century.

Both Planck and Einstein, the two greatest and most creative physicists of the past century,[1] were both realists and objectivists, as opposed to sensationists or purely subjectivists, as was Ernst Mach (1838–1916), who influenced both of them early in their scientific carriers and most of his followers. But Mach was soon left behind by Planck and Einstein when, forced both by their respective discoveries (the Theory of Quanta and the Theory of Relativity) they embraced realism as their philosophy of the physical world.

According to Planck,[1] speaking about the great pioneers: "What moved them (Copernicus, Kepler, Huygens, Newton and Faraday) was their firm belief in the reality of their picture (of the world) whether founded on an intellectual or on a religions basis." And, in another occasion, he wrote:

> As a firm starting point we are left with only a few principles, among them the universal law of energy.

9

Einstein[1] in a letter to his friend Maurice Solovine said:

> *You find it surprising that I think of the comprehensibility of the world (insofar as we are entitled to speak of such world) as a miracle or an eternal mystery. But, surely,* a priori, *one should expect the world to be chaotic, not to be grasped by thought in any way. One might (indeed should) expect that the world evidenced itself as lawful only so far as we grasp it in an orderly fashion. This would be a sort of order like the alphabetical order of words. On the other hand, the kind of order created, for example, by Newton's gravitational theory is of a very different character. Even if the axioms of the theory are posited by man, the success of such a procedure supposes in the objective world a degree of order which we are in no way entitled to expect* a priori.

In 1930 Wolfgang Pauli postulated the existence of the neutrino, a neutral particle of negligible mass, to justify the apparent loss of energy and linear momentum in the β disintegration process of neutrons:

$$n \rightarrow p + \bar{e} + \bar{\gamma}_e.$$

Being a neutral particle and having a negligible mass, if any, the neutrino would be extremely difficult to detect. Niels Bohr, who had early proposed the pioneering model for the hydrogen atom, and was then one of the most respected physicists in the world, did not believe in the neutrino's existence, and even went as far as saying that, perhaps, in the microscopic world of the atom, energy conservation was not an absolute requirement. This sounds a little strange in a man of his scientific caliber. But it was influential in discrediting for 25 years Pauli's proposal. In 1956, Clyde Cowan and Frederick Reines (who, many years later was awarded the Physics Nobel prize, which was also many years after the death of Cowan) demonstrated experimentally the existence of neutrinos bombarding pure water with a beam of 10^{18} neutrinos per second and subsequently detecting the photons emitted, in a very careful and laborious experiment: they had discovered the electron neutrino.

If needed, this was a most striking confirmation of energy conservation in the subatomic realm.

In 1987, Leo Max Lederman, Melvin Schwartz and Jack Steinberger discovered[2] the other two types of neutrinos, today known as the tau-neutrino and the muon-neutrino.

For some strange reason, in the physical realm of the whole cosmos, physicists have been careless with energy conservation. When, in 1948, Gold, Bondi and Hoyle[3] put forward their Steady State Model against the Big Bang Model, they had recourse to the *"continuous creation of matter"* in order to picture the general recession of galaxies in an unchanging *time-independent* universe. According to Hoyle[4] the idea of "continuous creation" can be described as "matter chasing its own tail", a fact in which he finds more "sweeping grandeur" than in the account of creation in Genesis, an account which is *not* meant, evidently to be a *scientific* description of anything, but which does not lack "grandeur" surely.

Apparently, not a few distinguished theoretical physicists, including some Nobel Prize winners, did not seem to care very much about energy conservation when, they did not appear to see anything wrong with the Steady State Theory.

Many years later, in 1980, Alan Guth[5] formulated his theory of cosmic inflation, which postulated an extraordinary (not conserving energy) cosmic expansion *at constant density*, taking place some 10^{-39} s after the Big Bang. As Guth says, the success of the Standard Model for elementary particles had produced a great impact within theoretical physicists. The fast pace of developments and the striking experimental confirmations by the late 70's resulted in the confident announcement that a full unification of all physics was at hand.

Grand Unification Theories (GUTs) predicted that the strong, weak and electromagnetic interactions would be unified at high energies ($E \approx 10^{16}$ GeV) and high temperatures ($T \approx E/k_B \approx 10^{29}$ K). These theories did not provide any hint as to why such high temperatures were needed to produce unification (symmetrization) of the interactions. Guth postulated a first-order (discontinuous) *phase transition* at still-higher temperatures ($T \approx 10^{31}$ K) at which the gravitational interaction would unify with the others. The phase transition would be analogous to the liquid–vapor transition, usually accompanied by supercooling, but quite different, in other respects, from phase transitions in the real world, which always conserve energy. According to Guth,[5] inflation (the ultimate in free lunch) takes place at constant density, between 10^{-37} s and 10^{-35} s, and results in an enormous growth in the cosmic radius, from $R = 10^{-52}$ m to $R \approx 2$ m.

In 1993, George Smoot gave an excellent talk at El Escorial on the recently measured CBR anisotropies detected by NASA's COBE satellite. He was talking at a Summer Course,[6] which I had organized together with Ignacio Cantarell and Rodolfo Nuñez de las Cuevas. At the end of George's talk, I made a comment: The equations in Inflationary Theory look very much like some equations in the old Steady State Theory resulting in growth at a constant density, but, now, conveniently back in time, some 10^{-37} s after the Big Bang, well beyond any possibility of being observed. His answer was that Inflation was not exactly the same as the Steady State Theory: a correct but perhaps insufficient answer.

Later, Vilenkin[7] summarize the story Alan Guth originally introduced his Inflationary Theory:

> *Guth was telling us that the veil of mystery surrounding the big bang could be finally lifted. His new theory would uncover the nature of the big bang and explain why the initial fireball was so-contrived. The seminar room fell suddenly silent. Everybody was intrigued...*
>
> *Everything reduces to this: "tunneling" from "nothing" to "something". A half physical, half metaphysical argument, deviced (apparently seriously) to dispense with the requirement of a free Supernatural Agent, creator of the cosmos, i.e., of a "natural" system ruled by "natural" laws.*
>
> *Apparently Guth and other cosmologists were looking for "marketable" ideas. And they succeeded. Like Dali and Picaso had succeeded before them.*
>
> *The Big Bang theory was increasingly identified thereafter with the end of inflation. "Finetuning" became instantly popular. The cosmological constant was identified with "the full energy density of vacuum". After a number of clever considerations "we conclude that quantum processes during inflation necessarily generate a distribution of regions of all possible values of the cosmological constant."*
>
> *And if the fundamental particle physics allows the constants to vary, then quantum processes during eternal inflation inescapably*

generate vast regions of space with all possible values of the constants. Eternal inflation thus provides a natural arena for applications of the anthropic principle.

In other words: too much "freedom".

It looks that the "freedom" denied to the Creator is somehow usurped by unscrupulous cosmologists in order to design literally countless imaginary worlds.

Quantum Mechanics is itself a "statistical" theory. Therefore it is not apt to describe a unique event, like the Big Bang. The Theory of Multiverses, so popular now for some time, is only a clever artificial concept, made up to jump from the real physical world we know to an imaginary multi-world existing only in our minds.

References

1. S. L. Jaki, *The Road of Science and the Ways to God.* (The University of Chicago Press, Chicago, 1978), Chapters 10 and 11, pp. 178, 192.
2. L. M. Lederman, *New Scientist*, 29 October 1988, p. 30.
3. H. Bondi, Spherical symmetrical models in general relativity, *Mon. Not. R. Astron. Soc.* **107** (1947) 410.
4. F. Hoyle, *The Nature of the Universe* (Harper Row, New York, 1950).
5. A. Guth, *The Inflationary Universe* (Perseus Books, Cambridge, Massachusetts, 1997), p. 187.
6. J. A. Gonzalo, J. L. Sanchez Gomez and M. A. Alario (eds.), *Cosmología Astrofísica* (Alianza Universidad, Madrid, 1995).
7. A. Vilenkin, *Many Worlds in One* (Hill & Wang, New York, 2006), pp. 10, 11, 137, 138.

Chapter 3

The Four Interactions

It is remarkable[1] that there are *four* and only *four* fundamental interactions in the universe. All stable physical entities from nuclei to galaxies are held together by a combination of the four interactions: the electromagnetic, the strong nuclear, the weak nuclear and the gravitational. Two decades ago, more or less, the *Physical Review Letters* published a number of papers claiming evidence for a *fifth* interaction, analogous to but different from the conventional gravitational interaction, which was finally discarded for lack of conclusive evidence.

In a solid for instance, *attractive* forces (of electromagnetic nature) must be present. Otherwise the solid would evaporate. And, at the same time, *repulsive* forces (also of electromagnetic nature) must be opposing the attractive forces to reach equilibrium. Otherwise the atoms making up the solid would collapse. The same type of equilibrium must be achieved in a star, like the Sun, between the enormous gravitational attraction pulling the individual particles toward the center of mass and the huge repulsion, due to radiation pressure generated by fusion at its core, pushing them away. In principle, one should say something similar about the elementary particles. If they have a finite size, as they do, the matter constituting the particle should be subject to competing attractive and repulsive interactions balancing each other, be they charged or neutral.

The *electromagnetic interaction* governs much of everyday life on our planet. All inorganic and organic chemical processes involve interactions between atoms, or ions, or molecular units which are of electric or magnetic character. Natural processes, like storms and hurricanes, are driven basically by electromagnetic interactions in the atmosphere, in the Earth's surface or in the oceans. Industrial processes, everyday life in large cities, transportation, communications, and agriculture, all depend to a large scale also on physical and chemical processes which are also driven by electromagnetic interactions.

The *strong nuclear* interaction is responsible for holding protons and neutrons together in atomic nuclei. It is strong enough to hold together many positively charged protons within heavy nuclei which repel each other strongly. The strong nuclear interaction acting among nucleons (protons and neutrons) is strong enough to bind uranium nuclei, with 92 protons, by adding 143 neutrons.

In spite of its short range, the strong interaction is therefore of the order of one hundred times stronger than the repulsive electromagnetic interaction. If it were one half weaker[2] the Periodic Table would contain substantially less stable elements, and life on Earth would not be possible. If it were twice as strong, the lifetime of midweight stars, like the Sun, would be much shorter.

The *weak nuclear* interaction, roughly 10^{-12} times weaker than the strong nuclear interaction, is important in many respects. For instance, it governs the spontaneous conversion of free neutrons in protons, and the inverse process. When a massive star explodes as a supernova,[2] the weak interaction is responsible for the energy deposited by neutrinos on the expanding shock front. The weak interaction precipitates the initial collapse, which allows the return of the heavier elements in the outer layer of the collapsing star to the interior of the galaxy.

In this way, for second or third generation stars like our Sun, their planets can have enough heavy elements essential for life. The weak interaction is also crucial in producing ^{4}He soon after the Big Bang, which plays an important role in the formation of stars.

The gravitational interaction, attractive always, by far the weakest at small scales, becomes dominant at large scales. The almost-negligible gravitational interaction, in large bodies (planets, stars, galaxies *etc.*), overwhelms completely the other interactions. Stars are formed out of galactic hot gas when gravity overcomes the gas pressure forces and the turbulence causing it to coalesce and contract. With a gravitational interaction 10^6 times stronger, according to Martin Rees[3]:

> *The number of atoms needed to make a star (a gravitationally bound fusion reactor) would be a billion times less... in this hypothetical strong-gravity world stellar lifetimes would be a million times shorter. Instead of living for ten billion years, a mini Sun would burn faster, and would have exhausted its energy before even the first steps in organic evolution had got under way.*

Such a star would have about one thousandth the luminosity, three times the surface temperature and one twentieth the density of the Sun, burning too hot and too quickly. The opposite would be true for a universe with a gravitational interaction substantially weaker.

Since the beginning of the twentieth century physicists begin to characterize the fundamental interactions in terms of *universal constants*: $\hbar$ (Planck), k_B (Boltzman), c (velocity of light), G (Newton), e (electron charge), m_e (electron mass), m_p (proton mass), m_n (neutron mass, slightly larger than proton mass), τ_n (free neutron half-life) and others, often combinations of the ones mentioned.

Planck noted that it is possible to define "natural" physical units by means of combinations of these universal constants. For instance

$$m_{\text{Pl}} = (\hbar c/G)^{1/2} = 2.17 \times 10^{-5} \, \text{g}, \tag{3.1}$$

$$l_{\text{Pl}} = (\hbar G/c^3)^{1/2} = 1.61 \times 10^{-33} \, \text{cm}, \tag{3.2}$$

$$t_{\text{Pl}} = (\hbar G/c^5)^{1/2} = 5.36 \times 10^{-44} \, \text{s}. \tag{3.3}$$

Table 3.1.   Cosmic quantities in Planck's units

Cosmic quantity	Natural unit	Ratio
$M_u = 1.547 \times 10^{54}$ g	$m_{\text{Pl}} = 2.17 \times 10^{-5}$	7.13×10^{58}
$R_+ = 4.588 \times 10^{26}$ cm	$l_{\text{Pl}} = 4.588 \times 10^{-33}$	2.85×10^{59}
$t_+ = 1.152 \times 10^{16}$ s	$t_{\text{Pl}} = 5.36 \times 10^{-44}$	2.15×10^{59}

The *relative intensities* of the fundamental interactions[4] can be given in terms of the following dimensionless quantities:

Gravitational interaction:

$$\alpha_G = \frac{Gm_p^2}{\hbar c} = 5.9 \times 10^{-39}$$

Electromagnetic interaction:

$$\alpha_E = \frac{e^2}{\hbar c} \approx \frac{1}{137}$$

Strong nuclear interaction[5]:

$$\alpha_{\text{NS}} = \frac{g_s^2 e^{-r/r'}}{\hbar c} \approx 1 \quad (g_S^2/\hbar c \approx 15; r' \approx 1.5 \times 10^{-13} \text{ cm})$$

Weak nuclear interaction[5]:

$$\alpha_{\text{NW}} = \frac{g_W^2 e^{-r/r''}}{\hbar c} \approx 10^{-12} \quad (g_w^2/\hbar c \approx 15 \times 10^{-12};$$

$$r'' \approx 1.5 \times 10^{-15} \text{ cm})$$

The dimensionless quantities introduced above are given for convenience in terms of the proton mass m_p and the charge e, but they could be given in terms of the Planck monopole's mass. For instance

$$\alpha_G(m_{\text{Pl}}) = \frac{Gm_{\text{Pl}}^2}{\hbar c} \approx 1, \tag{3.4}$$

where m_{Pl} is the massive hypothetical particle in which the gravitational interaction and the electromagnetic interaction are of the same order.

It is interesting to note that at the time stars and galaxies begun to form in the universe, at a cosmic radius $R_+ = 2GM_u/c^2$, the main cosmic quantities can be given in terms of Planck's natural units and the numbers for M_u (cosmic mass), R_+ (cosmic radius) and t_+ (cosmic time) come close to each other.

It may be noted that M_u (cosmic mass)/m_p (proton mass) gives the total number of baryons in the universe: 9.26×10^{80}.

Several physicists, including Dirac, Dicke, *etc.*, have considered the possibility of varying the universal constants with time through cosmic history. Up to now, the results have not been very conclusive.[6,7] In particular, the time-variation of Newton's gravitational constant, estimated from ancient eclipse records, and by various other means, is

$$(\dot{G}/G)_o \leq \text{between } 10^{-10} \text{ and } 10^{-11} \text{ per year;}$$

the time-variation of α, the electromagnetic fine-structure constant, estimated from measurements on uranium ores in the Oklo river (Gabon) is

$$(\dot{\alpha}/\alpha)_o \leq 10^{-6} \text{ per year,}$$

which implies either that e^2, $\hbar$ and c remain constant, or that they change simultaneously precisely to keep α almost constant, which does not seem very realistic.

References

1. R. Jastrow, *Red Giants and White Dwarfs* (Warner Books, New York, 1980), p. 46.
2. G. Gonzalez and J. W. Richards, *The Privileged Planet* (Reguery Publishing, Washington, 2004), pp. 201, 202.
3. M. Rees, *Just Six Numbers: The Deep Forces that Shape the Universe* (Basic Books, New York, 2000).
4. J. A. Gonzalo, *The Intelligible Universe* (World Scientific, Singapore, 2008).
5. R. Eisberg and R. Resnick, *Quantum Physics of Atoms, Molecules, Solids, Nuclei and Particles* (John Wiley: New York, 1947), pp. 689, 700.
6. F. Hoyle, *Frontiers of Astronomy* (London, Heinemann, 1955), p. 304.
7. H. Fritzsch, *The Fundamental Constant* (World Scientific, Singapore, 2009), p. 149.

Chapter 4

Matter, Radiation, Particles

Up to the beginning of the twentieth century, matter and radiation were considered as distinct entities. Matter, as making up a solid, a liquid, a gas or a plasma (a fluid of ionized atoms and electrons), was seen as altogether different from electromagnetic radiation, which spans from radio waves to X-rays and beyond. It may be noted that in a hot dense plasma, like the one making up a star, as our Sun, the electrons, which are almost two thousand times less massive than the protons or the ^{4}He nuclei, occupy much more volume than the much heavier protons and alpha particles. To begin with, the Compton wavelength of the electron,

$$\lambda_c(\text{electron}) = \frac{h}{m_e c^2} \approx 2.41 \times 10^{-10} \, \text{cm}, \tag{4.1}$$

is much larger than the Compton wavelength of the proton,

$$\lambda_c(\text{proton}) = \frac{h}{m_p c^2} \approx 1.31 \times 10^{-13} \, \text{cm}, \tag{4.2}$$

but one must take also into consideration Pauli's exclusion principle, which makes the effective electron radius in the plasma more like Bohr's radius[1]

$$a_0(\text{electron}) = \frac{\hbar^2}{m_e e^2} \approx 5.29 \times 10^{-9} \, \text{cm}, \tag{4.3}$$

21

almost twenty times larger than the electron Compton wavelength. This is true also for the plasma in a star and must be true too for the universe in its early plasma state, just after the primordial electrons could be formed, and well before decoupling (atom formation).

Arthur Holly Compton (1892–1962) discovered in 1923 that bombarding paraffin with X-rays they are scattered with a wavelength larger than the wavelength of the primary X-rays. This is called the Compton effect and, for its discovery, Compton was awarded, together with C. T. Wilson, the Physics Nobel Prize for 1927. The effect is not explicable with a purely undulatory conception of the electromagnetic radiation (X-rays or other), but it was explained, by Compton and by Peter Debye (later Chemistry Nobel Prize, in 1936), as due to the elastic collision between the radiation quanta (X-ray packages) and the electrons of the paraffin atoms. In these collisions energy and momentum were transferred from the radiation quanta (particles) to the electrons, respecting the principles of energy and momentum conservation.

Compton's views on the particle nature of radiation[1] were shared from the beginning by Maurice de Broglie, a French experimental physicist. His experiments and the analysis he made of them impressed very much his brother Louis, a student of history at the time, to such extent that Louis changed his career from history to physics. In 1924, Louis de Broglie presented his doctoral thesis to the Faculty of Science of the University of Paris arguing, originally and convincingly, in favor of the existence of matter waves. In essence, he was putting together Planck's ($E = \hbar\nu$) quantum theory and Einstein's relativity ($E = mc^2$). The originality of this thesis was immediately recognized, but, because of the lack of experimental evidence supporting it, was put aside for the moment by his physicist colleagues.

However, Albert Einstein recognized immediately its validity, and called the attention of other physicists to Louis de Broglie's work. In 1929 de Broglie was awarded for this discovery the Physics Nobel Prize.

Subatomic particles (not necessarily elementary) are characterized by a rest mass (m), a radius (r), a lifetime (τ), a charge (q),

Table 4.1. Subatomic particles

	m(g)	r(cm)	τ(s)	q(esu)	T(K)
Neutron	1.67×10^{-24}	1.2×10^{-13}	887	0	3.88×10^{12}
Proton	1.67×10^{-24}	1.2×10^{-13}	∞ (?)	4.8×10^{-10}	3.88×10^{12}
Electron	9.10×10^{-28}	(5.29×10^{-9})	∞ (?)	4.8×10^{-10}	2.11×10^{9}
Planck's monopole	2.17×10^{-28}	1.66×10^{-33}	—	—	5.04×10^{31}

a spin (intrinsic angular momentum) and a characteristic temperature $(T = mc^2/2.8k_B)$ above which they cannot exist as stable particles.

Experiments at high energy accelerators in the 1970's demonstrated unambiguously that neutrons and protons were made up by even more elementary constituents — the quarks. These elementary constituents, as far as we know, are not made up of smaller particles and have not internal structure.

According to the standard model there are three generations of *elementary fermions* (spin 1/2) including six quarks, three neutrinos, three leptons, their corresponding antiparticles, and four *elementary bosons* (spin 1) which mediate the electromagnetic (γ), the nuclear strong (g) and the nuclear weak (Z° and $W^\pm$) forces. The neutron, for instance, is made up of three quarks: d(down) d(down) u(up), and the proton of another set of three quarks: u(up) u(up) d(down). But the quarks themselves are not stable in isolation outside the nucleons of which they are part of. They are structureless and have a definite rest mass and a fractional $\pm$ charge. To have a definite physical existence, however fleeting, outside the nucleon, they must have some definite size, and some definite radius, which is difficult to measure and sometimes estimated as 10^{-18}cm, much smaller than the neutron or the proton radius.

What are the *cosmological implications*? A *material particle* with a given rest mass can be considered as a "pack" of energy tied up around itself, occupying a certain volume in space-time. One is therefore entitled to assign to it an average or characteristic radius and a characteristic density.

For instance, for a proton, its characteristic radius is

$$r_p(\text{proton}) \approx 1.2 \times 10^{-13}\,\text{cm} \tag{4.4}$$

and its characteristic density

$$\rho(\text{proton}) \approx \frac{m_p}{\frac{4\pi}{3}r_p^3} \approx 2.3 \times 10^{14}\,\text{g/cm}^3. \tag{4.5}$$

In the universe, at present, in our local neighborhood, we have galaxies, stars, planets and comic dust, made up of *atoms*, including nuclei and electrons, and a relatively small amount of radiation, the cosmic background radiation (CBR). Going back in time, the universe was denser and hotter, but the physical conditions were still apt to have atoms, which, before the formation of stars and galaxies, were in the form of dense and hot cosmic dust, surrounded by a hot cloud of radiation. Going further back in time, when the radiation temperature was around 3000–4000 degrees Kelvin, the atoms were ionized and the universe was a hot plasma of protons, ^{4}He nuclei and electrons, not yet too dense. At still earlier times the plasma was really hot: those were the times at which primordial nucleosynthesis took place. Cosmic density was then so high that individual electrons could not be separate entities. Before then, the universe contained protons (not yet ^{4}He nuclei) and some neutrons which had not yet disintegrated spontaneously. Still further back in time, at a temperature of about 3.88×10^{12} K, the density of the universe was nuclear density. Before that time cosmic density was larger than the density of ordinary neutrons or protons, so neither neutrons nor protons could exist as separate entities. Consequently, it is not very realistic to talk about a soup of nuclear particles and antiparticles at those times, because then they could not exist as such. Gamow, Alpher and Herman, spoke about the "ylem" (primordial matter) before the quarks were known to exist. They considered a moment in cosmic history at which the universe was an enormously dense aggregate of neutrons (so the net electric charge was zero) permeated by an enormous amount of radiation. This was their starting point to investigate the origin of nuclei heavier than hydrogen. Today we know that, leaving aside the nuclei of ^{4}He, and traces of other light elements, all other nuclei are cooked up in successive generations of stars. But this is another issue.

At earlier cosmic times that are hotter $(T > T = 3.88 \times 10^{12}\,\mathrm{K})$ it is highly speculative to reconstruct the thermal history of the universe. But one thing can be said. Before reaching the *singularity* $(t = 0, R = 0)$, Heisenberg's uncertainty principle for a universe with a finite mass,[2] say $M_u \approx 1.547 \times 10^{54}$ g, sets a limit to the time before which the speculation becomes meaningless.

We could call it Heisenberg's time

$$t_H \cong \frac{\hbar}{M_u c^2} \approx 7.54 \times 10^{-103}\,\mathrm{s}. \tag{4.6}$$

It may be noted that at these early times the cosmic radius $R(t)$ is given by Einstein's cosmological equation as

$$R(t) \approx \text{constant} \times t^{2/3}. \tag{4.7}$$

Then the growth of cosmic radius between $t = t_H$ (Heisenberg's time) and $t = t_{\mathrm{Pl}}$ (Planck's time) is given by

$$R(t_{\mathrm{Pl}})/R(t_H) = (t_{\mathrm{Pl}}/t_H)^{2/3} = 1.71 \times 10^{39}, \tag{4.8}$$

a number of the order of the inflationary factor used by Alan Guth $(\sim 10^{40})$, without any need of inflation.

References

1. R. Eisberg and R. Resnick, *Quantum Physics of Atoms, Molecules, Solids, Nuclei and Particles* (John Wiley, New York, 1974), pp. 63, 110.
2. J. A. Gonzalo, *Inflationary Cosmology Revisited* (World Scientific, Singapore, 2005), p. 55.

Chapter 5

The Dark Night Sky and Olbers' Paradox

Olber's paradox[1,2] is the riddle of the darkness of the night sky in conflict with the presumed infinity of the universe. If there were infinite stars, in infinite galaxies, in the universe, looking in any direction from the Earth's surface the line of sight should end at the surface of a star and then the whole night sky should look bright.

Since the German astronomer William Olbers (1758–1840) reformulated the dark night paradox in 1823 he and many after him have tried to save the presumption that the universe is infinite. Some did so for preconceived philosophical reasons, although not necessarily regarding Olbers. As Stanley L. Jaki points out[1] correctly in his carefully researched book, an infinite universe could readily pass for the ultimate entity, and serve thereby as a substitute for God.

According to R. H. Dicke (Princeton University): *"Professor Jaki has considered with great care the origins, history, and significance of this question and his scholarly, but interesting and readable book will be the definitive historical statement for years to come"*.

Jaki's book reprints, in the Appendices, two short essays by E. Halley,[3] one longer essay by J. P. Loÿs de Cheseaux[4] and Olbers' essay,[5] *On the Transparency of Space*. He also shows that Olbers' kept referring repeatedly to Halley's wrestling with the question, to such extent that the paradox could well have been named Halley's paradox.

The history of the ups and downs of Olbers' paradox through the nineteenth and twentieth centuries[1] is certainly illustrative of the fact that the awareness of the scientific community to key developments in physics does not grow with time properly and steadily at all. The "rediscovery" of Olbers' paradox in the mid twentieth century can be attributed to Bondi and Gold[6] (and somewhat later to Hoyle[7]) in connection with their Steady State Theory. Ten years later, in his Joule Memorial Lecture, H. Bondi makes the following claim:

> *In cosmology, which is the name given to the subject that investigates the large scale structure of the universe, we can put a precise date to the moment when it became a scientific subject and left the road of philosophical speculation. This date is 1826. In 1826, the German astronomer Olbers published a little investigation which, although I doubt whether he realized it, made cosmology a science.*

This inaccurate claim is no doubt exaggerated, but the importance of the question brought up by the recurrent paradox is certainly momentous.

Olbers, like most of his contemporaries, was convinced that the universe was infinite in extent and he gave some reasons, though not very satisfactory,[1] to show that the paradox of the darkness of the night sky could be somehow solved.

But the key to the solution of Olber's paradox would come with the Einstein General Theory of Relativity, through the assertion that the *total mass of the universe can only be finite*[8] if scientific considerations about the universe as a whole are to remain meaningful:

> *The total mass M of the universe, according to our view, is finite, and is in fact*

$$M = \rho \cdot 2\pi^2 R^3 = 4\pi^2 \frac{R}{k} = \pi^2 \left(\frac{32}{k^3 \rho} \right)^{1/2}. \qquad (5.1)$$

> *Thus the theoretical view of the actual universe, if it is in correspondence with our reasoning, is the following. The curvature of*

space is variable in time and place, according to the distribution of matter, but we may roughly approximate to it by means of a spherical space.

The total mass M of the universe could have been given perhaps more properly for a spherical cosmos by means of

$$M = \rho \cdot \frac{4\pi}{3} R^3, \tag{5.2}$$

with ρ as the average density, including mass matter and radiation matter, and R as the growing and now very large, but finite, radius of the universe. But the important point is, as Einstein points out boldly, that M should be *finite* to avoid physical inconsistencies.

Today we know that there are 10^{11} to 10^{12} galaxies, each containing 10^{11} to 10^{12} stars with an average mass of the other of the Sun's mass, so that $M_u \cong 1.54 \times 10^{54}$ g. The best telescopes in the world, at Hawaii, Chile, Tenerife *etc.*, could see galaxies substantially fainter than the faintest proto-galaxies or quasars *actually* observed by ground-based telescopes — but they do not *see* anything beyond a certain distance. The same is true of the space telescope Hubble. There are strong indications (through IR telescopes) that galaxies further away from the Milky Way are spatially more closely packed than galaxies in our immediate neighborhood. And the finiteness of the speed of light puts a limit to the recession velocity, and therefore to the maximum distance of observable protogalaxies emitting the light which is just arriving at us now, about thirteen billion years after they being emitted. As will be seen later the recession velocity of these young galaxies relative to us is beginning to saturate (it cannot exceed the speed of light relative to us) and one might interpret this as an effective acceleration of the galaxies in our neighborhood with respect to the most distant (youngest) galaxies.

It is possible to make a rough estimate of the ratio of radiation flux due to the CBR, which is at $T \approx 2.72$ K and is filling the cosmic plasma sphere as seen from our galaxy now, to that due to the visible light *if the whole sky* were fully occupied by stars in every direction.

The ratio of a nearby star's surface area at a distance

$$d_{\text{star}} \approx 5 \text{ light years} = 4.7 \times 10^{18} \text{ cm}$$

to the Sun's surface area at a distance

$$d_{sun} \approx 8 \text{ light minutes} = 1.44 \times 10^{13} cm$$

is the order of

$$\frac{A_{star}}{A_{sun}} = \left(\frac{d_{star}}{d_{sun}}\right)^2 \approx 9.56 \times 10^{-12}. \tag{5.3}$$

The Sun's surface temperature ($T_{sun} \approx 6000 \text{ K}$) is roughly equal to the surface temperature of a typical star, but the emitting surface, as seen from the Earth, is of course much larger. Then the estimate of the flux ratio would be

$$\frac{\Phi_{CBR}}{\Phi_{star}} = \frac{\Phi_{CBR}}{\Phi_{sun}}\left(\frac{A_{sun}}{A_{star}}\right) = \left(\frac{\sigma T_{CBR}^4}{\sigma T_{sun}^4}\right)\left(\frac{A_{sun}}{A_{star}}\right) \approx 4.41 \times 10^{-3}, \tag{5.4}$$

using Stefan–Boltzmann's law.

In a similar way, the ratio of the CBR flux to the flux of visible light coming from a night sky *if* filled with galaxies (distance between two typical nearby galaxies: $d_{gal} \approx 2 \times 10^6$ light years $= 1.89 \times 10^{24}$ cm), each containing about 10^{11} stars like our Sun, would be

$$\frac{\Phi_{CBR}}{\Phi_{gal}} = \frac{\Phi_{CBR}}{10^{11}\Phi_{sun}}\left(\frac{d_{gal}}{d_{sun}}\right)$$

$$= 10^{-11}\left(\frac{\sigma T_{CBR}^4}{\sigma T_{sun}^4}\right)\left(\frac{d_{gal}}{d_{sun}}\right)^2 \approx 2.09 \times 10^{-3}. \tag{5.5}$$

Of course, stars and galaxies do emit in the visible, but the CBR — a microwave radiation — is *invisible*, so that the night sky appears completely dark in spite of being pervaded by the very faint CBR radiation.

An observer looking to a night sky full of stars would see a star's surface in every direction of sight and would be seeing therefore a bright sky. The fact that we see the night sky as dark means therefore that the number of stars, and correspondingly the number of galaxies, is *finite*.

According to the Big Bang picture, just at t_H (Heisenberg time), R_H (Heisenberg radius), the universe was a tremendously concentrated *dot* of space-time filled with primordial *matter* and *radiation*.

Outside, as far as we can say, there was nothing: no matter, no radiation, no space, no time. Rather than undergoing an "explosion", the universe was undergoing an "implosion". The cosmic dot did grow *inside* the initial cosmic sphere. After a few microseconds, the first baryons (neutrons) did begin to exist; a few minutes later, the first electrons and protons appeared, massively, and neutrons and protons begun to fuse quickly into ^{4}He nuclei and other light nuclei in very small amounts; then nucleosynthesis stopped in about thirty minutes; much later, about a few million years later, atoms formed, the universe became transparent matter and radiation *decoupled* and the CBR began to flow from the outer space-time spherical shell towards the center. The Milky Way was moving then and now at a relatively small fraction of the speed of light with respect to the center of the CBR sphere. Therefore it can be considered to be placed relative near that center.

Much later, at about four hundred million years, the first stars and protogalaxies began to form. At that time the characteristic temperature the CBR was already as low as $T_+ \approx 60\,\mathrm{K}$, the cooling down and the expansion continued, and, finally, the Sun (a second or third generation star) formed, and around it the solar system, within which our planet Earth, about 4.8 billion years ago (i.e., about 8.9 billion years after the Big Bang).

At present, galaxies are some million years far away from each other, the CBR has cooled to $T \approx 2.72\,\mathrm{K}$, and the night sky looks dark. In the distant future, galaxies will be much farther apart, the CBR will be much cooler than it is now, and the night sky will look even darker.

As Eddington[9] said:

Infinity is a mischief and as such should have no place in physics.

Discussions of Olber's paradox in terms of the Steady State Theory, which assumed an infinite and eternal cosmos, received a death-blow in 1992, when the results of COBE (Cosmic Background Explorer) were made public.

In the early 80's Waldorn[10] (quoted by Jaki in *The Paradox of Olber's Paradox*) assumed, as apparently Halley had many years before, in order to solve the paradox, considered that as ever larger spherical shells are considered, the density of matter should decrease. This is exactly the opposite of what cosmic data in the IR support[11] and opposite of what Einstein's cosmological equations predict about cosmic evolution from early times to the present.

The *eternality* and *infinity* of matter is a fundamental Marxist dogma and as such it must have had, and still has considerable influence not only in the Communist world but also outside it.

J. Gribbin, in connection with Olber's paradox, makes the following comment:

> *The most fundamental observation in all of science is that night follows day. This simple fact is enough to show that the Universe has not always existed, everywhere, in the form that we see it today. There must be an "edge" to the Universe as we know it.*

The only way of removing Olber's paradox is to view it in terms of *cosmic finitude*. Attempts to do otherwise will end in inconsistent reasoning. Evidently, Olber's paradox cannot become a fundamental fact for cosmology when the universe becomes just our universe: one of many multiverses, which will be unobservable forever.

As Jaki[1] says:

> *It is not mathematics, but philosophy (and, one would add, common sense "physical realism", JAG) that alone conveys the impossibility of an actually realized infinite quantity and provides thereby the real foundation for a scientific respectable treatment of Olber's paradox.*

References

1. S. L. Jaki, *The Paradox of Olber's Paradox*, new edition (Real View Books, Pinckney, MI, 2000).
2. E. H. Harrison, *Darkness at Night. The Riddle of the Universe* (Harvard University Press, Harvard, Massachusetts, 1987).
3. E. Halley, *Philos. Trans.* **31** (1720) 22–26.
4. J. P. L. de Cheseaux, *Traité de la Comete* (Chez Marc-Michel Bousquet & Compagnie, Lausanne and Geneva, 1744).
5. W. Olbers, On the transparency of space, *Edinburgh New Philosophical Journal*, April–October 1826, pp. 141–150.
6. H. Bondi and T. Gold, *Mon. Not. Roy. Astron. Soc.* **108** (1948) 252–270.

7. F. Hoyle, *Frontiers of Astronomy* (Heinemann, London, 1955), p. 304.
8. A. Einstein, in *The Principle of Relativity: A Collection of Original Memoirs on the Special and General Theory of Relativity*, eds. H. A. Lorentz, A. Einstein, H. Perrett and G. B. Jeffrey (Dover, New York), pp. 177–188.
9. A. S. Eddington, *New Pathways in Science* (The University of Michigan Press, Ann Arbor, 1959), p. 217.
10. R. A. Waldorn, Is the Universe really expanding? *Speculations in Science and Technology* **4** (Dec. 1981) 539–543.
11. Private communication or, off hand remark to the author by R. A. Alpher, around 1993.

Part II: Relativistic Cosmology

6. General Relativity and Cosmology
7. The Friedmann–Lemaitre Solutions
8. The Role of Radiation Pressure
9. The Einstein–Lemaitre Correspondence

> *It is a magnificent feeling to recognize the unity*
> *of a complex of phenomena which appear to be things*
> *quite apart from the direct visible truth*

Albert Einstein
(1879–1955)

Chapter 6

General Relativity and Cosmology

The field equations of the General Theory of Relativity[1,2] are in principle the right *dynamical equations* to describe cosmic evolution. The field equations are given in terms of the Ricci tensor $R_{\mu\nu}$ (which is the contracted Riemann curvature tensor), the cosmological constant Λ (originally introduced by Einstein and finally discarded), and the stress-energy tensor $T_{\mu\nu}$ for all forms of matter and energy excluding gravity. $R_{\mu\nu}$ and R are determined by the *metric tensor* $g_{\mu\nu}$ together with its first and second-order partial derivatives. The metric tensor $g_{\mu\nu}$ is defined implicitly by

$$(ds)^2 = g_{\mu\nu}dx^\mu dx^\nu \tag{6.1}$$

under the assumption that space-time is locally Minkowskian (i.e., possessing the geometric properties of special relativity) and symmetric (i.e., invariant under the interchange of μ and ν), where x^μ and x^ν are four arbitrary coordinates describing fully the *curved space-time*.

The field equations are therefore a set of second-order nonlinear partial differential equations for $g_{\mu\nu}$. Initial and/or boundary conditions must be assumed in order to specify a unique solution.

Since the field equations are symmetric, there are in general ten independent equations, one for each independent $g_{\mu\nu}$. But, if the curved space-time is homogeneous and isotropic as it is assumed to be the case for the physical Universe, the number of independent equations reduces to two, and finally to one equation,

$$\dot{R}^2 = \frac{8\pi}{3}G\rho R^2 - kc^2 + \frac{\Lambda}{3}c^2 R^2 \qquad (6.2)$$

where R is the expanding cosmic radius, $\dot{R}$ its time derivative, G Newton's gravitational constant, ρ the mass density (of matter and radiation), k the space-time curvature, which can in principle be $k > 0$ (reinforcing gravity), $k = 0$ (neither reinforcing nor countering gravity), or $k < 0$ (countering gravity), c the speed of light in vacuum, and Λ the cosmological constant. Note that k is dimensionless, and Λ has the dimension of an inverse-square length. As mentioned, Λ was originally introduced by Einstein to counter gravity in order to get a static universe.

This probably means that the author of general relativity was thinking then in a $k > 0$, reinforcing gravity. When Friedmann[3] found the correct solutions for a *spherical homogeneous* and *isotropic* universe, he assumed $\Lambda = 0$, and saw that $k < 0$ had in fact an effect opposite to gravity, resulting in an expanding universe.

The original theory of gravitation, as it is well known, was formulated by Newton implying an instantaneous interaction between massive bodies. This is equivalent to assuming that the gravitational interaction propagates at infinite speed. To correct this shortcoming, Einstein formulated in 1916 his General Theory of Relativity, treating the gravitational interaction in a fully consistent way. The *principle of equivalence* says that an observer cannot distinguish between being in a system of reference which is subject to a gravitational field and being in one which is uniformly accelerated. This is tantamount to say that *inertial* mass and *gravitational* mass are equal. Physical laws must be such, according to Einstein, that they apply to any reference system regardless of its motion with respect to the mass distribution in the universe. Einstein took recourse to *curved spaces*, as introduced first by Gauss, Bolyai and Lobachevski (for the case of a *two-dimensional* space of constant curvature) and later generalized by Riemann (for *more dimensions*), to develop the complex mathematical formalism that is appropriate to describe relativistically the dynamics of accelerated systems.

But (and this is very important) *if* we take into account that the actual geometry of the universe is *spherical*, and that the cosmic distribution of matter (galaxies) and radiation (the CBR) is basically *homogeneous* and *isotropic*, the geometry of cosmic space-time reduces to that of a set of two-dimensional surfaces describable by a single curvature parameter K, which can be

$$K = \begin{cases} \dfrac{1}{a^2} \text{ (closed spherical surfaces with } positive \text{ curvature)} \\ 0 \text{ (plane, Euclidean space)} \\ -\dfrac{1}{a^2} \text{ (open Gauss, Bolyai, Lobachevski surfaces with} \\ \quad negative \text{ curvature),} \end{cases}$$

where, if a (the scale factor) satisfies $a \gg 1$, the closed and open symmetries became indistinguishable in practice.

Note that Einstein's cosmological equation Eq. (6.2) involves only one space-time curvature k, plus the artificial, and in fact unnecessary, counter-gravitational term Λ. Note also that the integration of Eq. (6.2) becomes insensitive to the values of the curvature $|k|$, for

$$R \ll R_+ = 2GM/|k|c^2$$

Previous to his application of the General Theory of Relativity to describe cosmic dynamics, Einstein had applied it successfully to predict such small gravitational effects as the bending of light rays in the Sun's gravitational field, the advance in the perihelion of mercury and the gravitational redshift of the light emitted from very massive objects. In all these cases the minor corrections to Newtonian gravity involved always *positively* curved distortions ($k > 0$) in the space-time. So the possibility of *negatively* curved distortions ($k < 0$) were probably not contemplated at this stage.

Only after Friedmann worked out systematically the set of all possible solutions (with $\Lambda = 0$) the possibility of a negatively curved space, involving anti-gravitational effects, was realized. Note that at the time the discovery of the CBR (and the possibility of cosmic

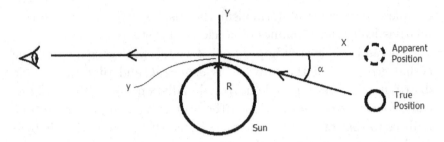

Fig. 6.1.   Deviation of light ray from a star when it passes near the Sun's surface: $\alpha \approx sGM/Rc^2$ where s is a dimensionless factor of order unity, G Newton's constant, M the Sun's mass, R the Sun's radius and c the speed of light.

radiation pressure effects) was yet to come many years into the future.

Let us consider[4] the bending of light rays by the Sun's gravitational field.

If a material particle or a photon with mass m moves in the vicinity of a body of large mass M, like the Sun, the first-order corrections to Newton's gravitational potential may be expected to be of the order of

$$E_p/mc^2 = (GMm/r)/mc^2 = GM/c^2r, \qquad (6.3)$$

which is independent of m, the mass of the particle. Let us assume that the observed intervals of space and time, under the gravitational distortion due to M, are given, in a first approximation, by

$$\Delta x' \approx \left(1 - a\frac{GM}{c^2r}\right)\Delta x, \qquad (6.4)$$

$$\Delta t' \approx \left(1 - b\frac{GM}{c^2r}\right)\Delta t, \qquad (6.5)$$

where a and b are dimensionless constants of order unity.

The velocity of the moving particle, under these assumptions, would be given by

$$v' \approx \frac{\left(1 - a\frac{GM}{c^2r}\right)}{\left(1 - b\frac{GM}{c^2r}\right)}v = \left(1 - s\frac{GM}{c^2r}\right)v \qquad (6.6)$$

where $s = a - b$ should be also of order unity. In particular, for the velocity of light under the influence of the strong gravitational field, one would have

$$c' = \left(1 - s\frac{GM}{c^2 r}\right) c. \tag{6.7}$$

To estimate the deviation of the light ray by the gravitational field of the Sun we set

$$d\alpha = \frac{dx' - dx}{dy} = \left(\frac{dc'}{dy}\right) dt = \frac{1}{c'}\left(\frac{dc'}{dy}\right) dx, \tag{6.8}$$

where c' is given by Eq. (6.7). Here

$$\frac{dc'}{dy} = s\frac{GM}{c^2 r}\frac{dr}{dy}, \tag{6.9}$$

and for a ray tangent to the Sun's surface $\left(y = 0, \ \frac{dr}{dy} = \frac{R}{r}\right)$

$$\frac{dc'}{dy} = \frac{2GMR}{cr^3}. \tag{6.10}$$

Therefore

$$\Delta\alpha \approx \frac{sGMR}{c^2} \int_{-\infty}^{\infty} \frac{dx}{r^3} = \frac{sGMR}{c^2}\left[\frac{2}{R^2}\right] = \frac{2sGM}{c^2 R}. \tag{6.11}$$

Substituting numerical data

$$G = 6.67 \times 10^{-8} \text{ (cgs units)}$$
$$M = 2 \times 10^{33} \text{ g}$$
$$c = 3 \times 10^{10} \text{ cm/s}$$
$$R = 6.96 \times 10^{10} \text{ cm}$$

we get, assuming $s \approx 1$,

$$\Delta\alpha \approx 2.1 \times 10^{-6} \text{ rad} \approx 0.43 \text{ arcsecond},$$

which is a small but observable quantity.

But the most ambitious application of the General Theory of Relativity was made by Einstein himself when he applied it to the entire observable universe. And he concluded that, to avoid physical (not necessary mathematical) inconsistencies the universe must be finite.

Note that, as pointed out by Stanley L. Jaki,[5] Newton preferred to use the expression "an immense universe" instead of "an infinite" universe but, from his correspondence with Bentley, it appears that he argued in favor of a universe with infinite stars, overlooking the fact that a universe with infinite stars homogeneously distributed would give rise to infinite local gravitational forces pulling apart any material object in every direction.

In a closed universe (which would have a maximum finite space-time volume) extending a line element along a geodesic (the equivalent to a straight line in a Euclidean space) would end up coming back to the same point where it had begun. In an open universe (which has no boundaries, i.e., can grow indefinitely in space-time remaining finite) extending one line element along a geodesic would never come back to the point of departure. This is what would happen to a light ray propagating through either curved space. And also to material particles, or, for that matter, to very massive galaxies, moving in the respective local cosmic geometry.

In an open universe the particles, or galaxies, would not only move under the influence of the cosmic mass within the sphere interior to the point where the particle (or galaxy) was, *but* also, due to the peculiar geometry of the open space-time, they would notice a *counter-gravitational* force which could be in principle like a cosmic radiation pressure, like the one associated to the Cosmic Background Radiation at that particular point in space-time. As the cosmic radius R increases, the corresponding cosmic temperature T (associated with the CBR) decreases, and so does the cosmic radiation pressure, approaching zero but never reaching zero in a finite time.

The first one to take seriously the idea of an *open universe* ($k < 0$) was the Russian mathematician Alexander Friedmann[3] in 1922. The second one, independently, was the former World War I artillery officer (and later Catholic priest) Georges Lemaitre[6] in 1927. Lemaitre connected his open solutions to Einstein's equation with the *general expansion* of galaxies in the universe being observed at that time in the Mount Wilson Observatory by E. Hubble and M. L. Humason. He was the first to infer a primeval explosion of matter and energy out of a tremendously dense and hot *primeval atom*, introducing in this way what was latter called the Big Bang model of the origin of the universe.

References

1. S. Weinberg, *Gravitation and Cosmology* (John Wiley and Sons, New York, 1972).
2. S. Weinberg, *Cosmology* (Oxford University Press, Oxford, 2008).
3. A. Friedmann, *Z. Phys.* **10** (1922) 377.
4. M. Alonso and E. J. Finn, *Physics* (Addison-Wesley, Reading, Massachusetts, 1992).
5. S. L. Jaki, *The Road of Science and the Ways to God* (The Universe of Chicago Press, Chicago, 1978).
6. G. Lemaitre, *Ann. Soc. Sci. Brux.* **A47** (1927) 49.

Chapter 7

The Friedmann–Lemaitre Solutions

As we have anticipated in Chapter 1, Eq. (1.1), and discussed further in Chapter 6, Eq. (6.2), Einstein's cosmological equation is

$$\dot{R}^2 = \frac{8\pi}{3}G\rho R^2 - kc^2 + \frac{\Lambda}{3}c^2 R^2, \tag{7.1}$$

where R is the cosmic radius, $\dot{R}$ its time derivative, G Newton's gravitational constant ($G = 6.67 \times 10^{-8}$ in cgs units), ρ the mass density (including matter mass and radiation mass), k the dimensionless space-time curvature, c the speed of light in vacuum ($c = 3 \times 10^{10}$ cm/s), and Λ the Einstein's cosmological constant.

We have mentioned previously that the term involving the cosmological constant becomes equivalent to the term involving the space-time curvature by indentifying $\Lambda = -3k^2/R^2$. So, assuming $k = 0$, $\Lambda \neq 0$ is fully equivalent to assuming $k = -|k| \neq 0$, $\Lambda = 0$. For a spherical universe of finite mass and open ($k < 0$) space-time geometry, Eq. (7.1) becomes

$$\dot{R}^2 = \frac{2GM}{R} + |k|c^2, \tag{7.2}$$

where $M = \frac{4\pi}{3}R^3\rho$ is the total constant mass of the universe.

It is convenient to define a characteristic cosmic radius R_+ such that

$$\frac{2GM}{R_+} = |k|c^2, \quad R_+ = \frac{2GM}{|k|c^2}. \tag{7.3}$$

This radius is directly related to the cosmic Schwarzschild radius for a universe with mass M and sets the length scale for the universe.

Note that for $R \ll R_+$ the second term in Eq. (7.1) can be neglected and the universe's space-time geometry becomes indistinguishable from Euclidean ($k = 0$). For $R \gg R_+$ on the other hand the first term can be neglected, and the speed at which the cosmic radius grows approaches $\dot{R} \approx |k|^{1/2}c$. Note that if $M \to \infty$, $R_+ \to \infty$, and then the space-time geometry remains Euclidean at any $R < R_+$. Precisely because a universe with an infinite mass is irremediably paradoxical, cosmic space-time cannot be Euclidean.

Thus we are justified to assume that the universe has a finite mass and is open.

Lemaitre, considered the father of Big Bang cosmology, describes[1] the beginning of the universe in the following words:

> *The atomic world broke into fragments and each fragment in smaller bits. Assuming, for simplicity, that this fragmentation took place in bits of equal size, two hundred and sixty successive fragmentations must have taken place so as to achieve the fragmentation necessary to reach the fragmentation of our poor little atoms, so small that they can hardly be broken further. The world evolution can be compared to a just finished fireworks event: a few extinguishing red fires, ashes and smoke. Facing a cooling red hot coal, we see the slow end of the suns and try to imagine the vanishing brightness of the origin of the worlds.*

About twenty years later, in November 22, 1951, Pope Pius XII addressed the Pontificia Academia of Sciences, of which Lemaitre was then president, suggesting a certain amount of agreement between the original condition of the first matter of the universe according to some current scientific theories, and the august instant of the "Fiat Lux".

Fred Hoyle,[2] a staunch supporter of the Steady State Theory, then a dignified rival of the Big Bang, said "What kind of scientific theory is this that was conceived by a priest and endorsed by a pope?"

For years the Big Bang theory was treated with suspicion, especially because the age of the universe deduced by Lemaitre from Hubble's estimate of $\dot{R}/R$ was much less than the current Earth's age determined by means of radioactive dating. This was so because the cosmic distance scale used by Hubble was in error by a factor of ten.

Only after the discovery[3] of the cosmic background radiation by Penzias and Wilson the Big Bang theory, at first reluctantly, then enthusiastically begun to be accepted by the scientific community at large.

Let us go back to Einstein's cosmological compact Friedmann–Lemaitre solutions in terms of R_+. They are obtained as follows: Eq. (7.1) for an open universe is rewritten as

$$\dot{R}^2 = R^{-1/2} \left\{ \frac{8\pi G}{3} \rho R^3 + c^2 |k| R \right\}^{1/2} \tag{7.4}$$

which can be integrated as

$$\int dt = \int \frac{R^{1/2}}{\{2GM + c^2 |k| R\}^{1/2}} dR. \tag{7.5}$$

Using the change of variable

$$x^2 = c^2 |k| R \tag{7.6}$$

it can be rewritten as

$$\int \frac{x^2}{(a^2 + x^2)^{1/2}} dx \tag{7.7}$$

which can be found in tables, resulting in

$$t = \frac{R_+}{c|k|^{1/2}} [\sinh(y) \cosh(y) - y], \tag{7.8}$$

where $y \equiv \sinh^{-1}(R/R_+)^{1/2}$, which implies

$$R = R_+ \sinh^2 y. \tag{7.9}$$

From the compact parametric solutions $t(y)$, $R(y)$, in terms of $R_+ = 2GM/|k|c^2$, we can get easily

$$\dot{R} = |k|^{1/2}c/\tanh y, \tag{7.10}$$

$$\ddot{R} = -\frac{1}{2}\frac{|k|^{1/2}c}{R_+} \bigg/ \sinh^4 y, \tag{7.11}$$

$$H = \dot{R}/R = \frac{|k|^{1/2}c}{R_+}\cosh y \bigg/ \sinh^3 y, \tag{7.12}$$

$$\rho/\rho_c \equiv \Omega = 1/\cosh^2 y = 1 - \tanh^2 y, \tag{7.13}$$

$$-\ddot{R}R/\dot{R}^2 \equiv q = \frac{1}{2|k|^{1/2}}(1/\cosh^2 y). \tag{7.14}$$

In particular, the dimensionless product Ht is given by

$$Ht = \frac{[\sinh(y)\cosh(y) - y]\cosh(y)}{\sinh^3(y)} \tag{7.15}$$

and therefore $2/3 \leq Ht \leq 1$ for y going from 0 to $\gg 1$.

It may be noted that using the present observational data[4] for $H_o = 67\,\text{km/sMpc} = 2.18 \times 10^{-18}\,\text{s}^{-1}$ (within 5.6%) and to $13.7\,\text{Gyrs} = 4.32 \times 10^{17}\,\text{s}$ (within 1.5%) the product

$$H_o t_o = 0.942 \pm 0.065 \tag{7.16}$$

is incompatible with a closed universe ($k > 0$), which implies $2/3 \geq Ht \geq 0$.

On the other hand, the dimensionless ratio $\Omega = \rho/\rho_c$, where $\rho = \rho_m + \rho_r$ is the sum of the matter mass density and the radiation mass density, and where $\rho_c = 3H^2/8\pi G$ is the critical mass density, is given by

$$\Omega = \Omega_m + \Omega_r = \frac{1}{\cosh^2(y)}1 - \tanh^2(y), \tag{7.17}$$

and therefore $1 \leq \Omega \leq 0$ for y going from 0 to $\gg 1$.

Note that the present observational values for H_o and t_o fix the present value of y_o through

$$H_o t_o = \frac{[\sinh(y_o)\cosh(y_o) - y_o]\cosh(y_o)}{\sinh^3(y_o)} = 0.942 \pm 0.065 \tag{7.18}$$

resulting in

$$y_o = \sinh^{-1}(R_o/R_+) = \sinh^{-1}(T_+/T_o) = 2.243 \pm 0.157 \quad (7.19)$$

where the equation of state $RT = R_oT_o = $ constant (the present transparent phase of the universe) has been used. Taking into account that $R = R_+$ (Schwarzschild radius), $\dot{R}_+ = |k|c^2/\tanh(y) \approx c$, where $\sinh^{-1}(y_+) = 1$, resulting in $|k| = 1/2$, and Eq. (7.8) for $y = y_o = 2.243$, $t(y) = t(y_o)$ results in

$$R_+ = 4.588 \times 10^{26} \text{ cm}$$
$$(R_o = R_+ \sinh^2 y_o = 9.96 \times 10^{27} \text{ cm}),$$

and, correspondingly, that

$$T_+ = (R_o/R_+)T_o = (21.69)2.726 = 59.2 \text{ K},$$

the set of principal cosmic numbers being sufficiently specified. Using $y_o = 2.243$, the dimensionless ratio

$$\Omega_o = \Omega_{mo} + \Omega_{ro} = 1 - \tanh^2 y_o = 0.044 \quad (7.20)$$

is obtained. This is again incompatible with a closed universe ($k > 0$), which implies $1 \le \Omega$.

Knowing numerically R_+ and $|k|$ the finite total matter mass of the universe can be evaluated as being

$$M = \frac{R_+c^2}{2G|k|^{1/2}} = 1.547 \times 10^{54} \text{ g.} \quad (7.21)$$

Dividing Einstein's cosmological equation (7.1) with $k < 0$, $\Lambda = 0$, by $\frac{8\pi}{3}GR^2$ we get

$$\frac{3(\dot{R}/R)^2}{8\pi G} = \rho + \frac{3(\dot{R}/R)^2}{8\pi G}|k|\left(\frac{c}{\dot{R}}\right)^2, \quad (7.22)$$

and, taking into account that $\frac{3(\dot{R}/R)^2}{8\pi G} = \rho_c$ (the critical density), we get

$$1 = \left(\frac{\rho}{\rho_c}\right) + |k|\left(\frac{c}{\dot{R}}\right)^2 = (\Omega_m + \Omega_r) + \Omega_k, \quad (7.23)$$

where we have defined Ω_k as $\Omega_k = |k|(c/\dot{R})^2$. If we identify the left-hand side of Eq. (7.1) with the *kinetic energy* and the first term in the right-hand side Eq. (7.1) with the *gravitational potential energy*, we could identify the second term Ω_k as *the space-time curvature potential energy*. Then

$$\Omega = \frac{1}{\cosh^2 y} = 1 - \tanh^2 y, \tag{7.24}$$

$$\Omega_k = 1 - \Omega = \tanh^2 y. \tag{7.25}$$

To separate the mass matter and mass radiation contribution in Ω we make use of the equation of state for the transparent phase of the universe's expansion $(R \geq R_{af}, T \leq T_{af})$, where R_{af} and T_{af} are the radius and CBR temperature at decoupling $(\rho_{maf} = \rho_{raf}$, atom formation)

$$RT = R_{af}T_{af} = \text{constant}, \tag{7.26}$$

and also of the fact that

$$\rho_m(R) = \rho_{maf}\left(\frac{R_{af}}{R}\right)^3, \left[\rho_m(R) = \frac{M}{\frac{4\pi}{3}R^3}\right], \tag{7.27}$$

$$\rho_r(R) = \rho_{raf}\left(\frac{T}{T_{af}}\right)^4, \left[\rho_r(T) = \sigma T^4\right], \tag{7.28}$$

to get

$$\frac{\rho_r}{\rho_m} = \left(\frac{RT}{R_{af}T_{af}}\right)^3 \frac{T}{T_{af}} = \frac{T}{T_{af}}. \tag{7.29}$$

Then

$$\Omega_m = \Omega \bigg/ \left(1 + \frac{T}{T_{af}}\right), \tag{7.30}$$

$$\Omega_r = \Omega \left(\frac{T}{T_{af}}\right) \bigg/ \left(1 + \frac{T}{T_{af}}\right). \tag{7.31}$$

Here T (decoupling) $= T_{af} = 2968\,\text{K}$ and $T_+ = 59.2\,\text{K}$ (as pointed out above for $y_o = 2.243$).[5]

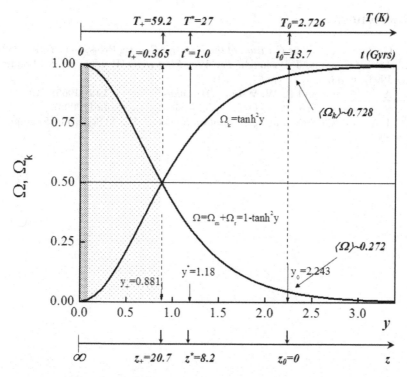

Fig. 7.1. Contributions to the density parameter Ω_m (matter), Ω_r (radiation), Ω_k (space-time curvature) as a function of y (dimensionless cosmic parameter), z (redshift), t (time) and T (CBR temperature).

Figure 7.1 displays the evolution of Ω (matter plus radiation), and Ω_k (space-time curvature) as a function of $T(\mathrm{K})$, $t(\mathrm{Gyrs})$, y (dimensionless cosmic parameter), and z (redshift) from $y = 0$ to $y \approx 3.5$, which spans the early plasma event at which atoms form ($T_{\mathrm{af}} = 2968\,\mathrm{K}$), the moment ($\Omega = \Omega_k$) at which stars and galaxies begin to form ($T_+ = 59.2\,\mathrm{K}$), and the present epoch ($T = 2.726\,\mathrm{K}$).

The universe as we see it now encompasses from protogalaxies and protostars formed at about $T^* \approx 27\,\mathrm{K}$, $t^* \approx 1\,\mathrm{Gyr}$, $y^* \approx 1.18$, $z^* \approx 8.2$ to galaxies and stars formed more recently as we see them now at $T_o \approx 2.726\,\mathrm{K}$, $t_o \approx 13.7\,\mathrm{Gyrs}$, $y_o \approx 2.243$, $z_o \approx 0$.

References

1. G. Gammow, in *The Creation of the Universe* (Viking Press, New York, 1952).
2. J. C. Mather and J. Boslough, *The Very First Light* (Penguin Books, London, 1996), p. 48.
3. A. A. Penzias and R. W. Wilson, *Astrophys. Journal*, **142** (1965) 419.
4. S. Weinberg, *Cosmology* (Oxford University Press, Oxford, 2008).
5. J. A. Gonzalo, *Inflationary Cosmology Revisited* (World Scientific, Singapore, 2005).

Chapter 8

The Role of Radiation Pressure

Alan Lightman and Roberta Brawer in *Origins*[1] (a set of interviews with leading cosmologists giving their perspective in 1990) suggest the year 1965 — the year in which Arno Penzias and Robert Wilson reported their discovery of the *cosmic background radiation* (CBR) — as the starting year of scientific cosmology.

This suggestion may be a little exaggerated. But, as Ralph Alpher and Robert Herman — the predictors of the CBR in 1948 — note in their latest book,[2] the year 1965 was a pivotal year for scientific cosmology. The importance of the cosmic radiation, which had been more or less overlooked by Einstein, Friedmann, Lemaitre, Eddington, Hubble and other pioneer cosmologists, would be certainly decisive in bringing to maturity modern scientific cosmology.

In 1964, satellite communications was entering a period of rapid development. Researchers at leading laboratories were improving their equipment to detect clean signals for commercial exploitation of the new technology. That year A. Penzias and R. Wilson, at Bell Labs, were developing a microwave receiver with a very low noise figure, with the aim of using it as a receiver for satellite communications. As it is well known, after eliminating carefully all conceivable sources of unwanted noise in their antenna, they were left with a persistent strange signal. It was the background radiation, which in the

Fig. 8.1.   Planck's blackbody radiation spectrum giving the relative (monochromatic) brightness in ergs/cm²·sec as a function of the wavelength in cm. The CBR observed by Penzias and Wilson in 1965 at 7.3 cm corresponded to a characteristic temperature of about 3.5 K.

present cosmic epoch is at a wavelength of 7.3 cm in the microwave region of the electromagnetic spectrum. The persistent background radiation appeared to be unpolarized and isotropic, independent of the direction of observation, and with an equivalent temperature of about 3.5 K if the source was assumed to be blackbody, as later confirmed. The blackbody character of the radiation, as shown by Max Planck in 1900, was completely specified by a single number: the value of the equivalent temperature. (See Fig. 8.1.)

In his 1965 paper[3] Penzias and Wilson reported that, at first, they did not know they were making a fundamental cosmological observation but that was precisely what it was. Given the blackbody character of the CBR, one measurement at a single wavelength was sufficient to characterize the spectrum in the full spectral range.

Let us describe briefly the role played by the cosmic background radiation through the expansion.

In its early *plasma phase* (at $T > T_{\text{decoupling}} \approx T_{\text{atom formation}}$) the universe was opaque, and matter and radiation had to expand in unison. In fact radiation pressure was pushing matter away radially (i.e., material particles, after baryon formation, electron formation and primordial nucleosynthesis). In this earlier *plasma* phase

$$\rho = \frac{M}{\frac{4\pi}{3}R^3} \text{ (matter mass density)}, \tag{8.1}$$

and

$$\rho = \frac{\sigma T^4}{c^2} \text{ (radiation mass density)}, \tag{8.2}$$

according to the Stefan–Boltzmann law for blackbody radiation. The cosmic equation of state before decoupling (atom formation) was therefore

$$\rho_r/\rho_m = \rho_{\text{raf}}/\rho_{\text{maf}} \rightarrow T^4 R^3 = T_{\text{af}}^4 R_{\text{af}}^3 = \text{constant } (t \leq t_{\text{af}}). \tag{8.3}$$

Then, at $t = t_{\text{af}}$ the cosmic temperature had decreased sufficiently to let electrons and protons (or ^{4}He nuclei) form hydrogen and helium atoms. The universe became transparent, and radiation propagated much more freely; because in a transparent universe the total number of photons is conserved, as is the total number of baryons, we have

$$n_r/n_b = \left(\frac{\sigma T^4}{2.8\,k_B T}\right)\bigg/\left(\frac{1}{m_b}\frac{M}{4\pi R^3/3}\right)$$

$$= \text{constant} \times T^3 R^3 = \text{constant} \tag{8.4}$$

resulting in a different equation of state

$$T^3 R^3 = T_{\text{af}}^3 R_{\text{af}}^3 = \text{constant } (t \geq t_{\text{af}}). \tag{8.5}$$

Note that the ratio ρ_r/ρ_m is no longer constant (and equal to unity) at $t \geq t_{\text{af}}$ but becomes instead

$$\rho_r/\rho_m = (\rho_{\text{raf}}/\rho_{\text{maf}}) \cdot (TR/T_{\text{af}}R_{\text{af}})^3 \cdot (T/T_{\text{af}}) = T/T_{\text{af}}. \tag{8.6}$$

We can estimate the *ratio* of radiation pressure

$$p_r = \frac{1}{3}\sigma T^4$$

to matter pressure

$$p_m = \frac{nk_BT}{V} = \left(\frac{nm_b}{V}\right)\frac{k_BT}{m_b} = \rho_m\frac{k_BT}{m_b},$$

where n is the number of baryons per unit volume and m_b the baryon mass. Then

$$\frac{p_r}{p_m} = \frac{\frac{1}{3}\rho_r c^2}{\rho_m\frac{k_BT}{m_b}} = \frac{1}{3}\left(\frac{T}{T_{\text{af}}}\right)\frac{m_b c^2}{k_BT} = \frac{1}{3}\frac{m_b c^2}{k_BT_{\text{af}}} \approx 9.35 \times 10^8. \qquad (8.7)$$

In other words, the *radiation pressure* in the present transparent phase of cosmic expansion is much larger than the *matter pressure*, itself larger than the gravitational attraction. This is in consonance with the *open* $(k < 0)$ space-time character of our universe.

We can make a direct comparison among the ways k affects cosmic space-time for the three cases of an *open* $(k < 0)$, a Euclidean or *flat* $(k = 0)$, and a *closed* $(k > 0)$ universe. For an *open* universe, as we have seen

$$R = R_+\sinh^2 y, \quad t = \frac{R_+}{|k|^{1/2}c}\{\sinh y\cosh y - y\}, \qquad (8.8)$$

where

$$\sinh y \approx y + \frac{y^3}{3!} + \frac{y^5}{5!}\ldots, \quad \cosh y \approx 1 + \frac{y^2}{2!} + \frac{y^4}{4!} + \frac{y^6}{6!}\ldots$$

For a *flat* universe we can interpolate

$$R = R_+ y^2, \quad t = \frac{2}{3}\frac{R_+}{c}y^3. \qquad (8.9)$$

For a *closed* universe,

$$R = R_+\sin^2 y, \quad t = \frac{R_+}{|k|^{1/2}c}\{y - \sin y\cos y\}, \qquad (8.10)$$

where

$$\sin y \approx y - \frac{y^3}{3!} + \frac{y^5}{5!}\ldots, \quad \cos y \approx 1 - \frac{y^2}{2!} + \frac{y^4}{4!} - \frac{y^6}{6!}\ldots$$

Therefore, in an *open* $(k < 0)$ universe, space and time are both dilated with respect to the *flat* $(k = 0)$ universe:

$$\frac{[R(y)]_{k<0}}{[R(y)]_{k=0}} > 1, \quad \frac{[t(y)]_{k<0}}{[t(y)]_{k=0}} > 1.$$

And in a *closed* ($k > 0$) universe, space and time are both *contracted* with respect to the flat ($k = 0$) universe

$$\frac{[R(y)]_{k>0}}{[R(y)]_{k=0}} < 1, \quad \frac{[t(y)]_{k>0}}{[t(y)]_{k=0}} < 1.$$

The growth speed $\dot{R}$ of the cosmic radius at $y \ll 1$ is therefore given by

$$\dot{R} = \frac{dR/dy}{dt/dy} = c(|k|^{1/2}/y)[1 + sy^2 + \cdots] \text{ for } k < 0,$$

$$\dot{R} = \frac{dR/dy}{dt/dy} = c(1/y) \quad \text{for } k = 0, \text{ and}$$

$$\dot{R} = \frac{dR/dy}{dt/dy} = c(|k|^{1/2}/y)[1 - sy^2 + \cdots] \text{ for } k > 0,$$

with s a numerical factor of order unitary. Note that at $y \ll 1$, $\dot{R} \gg c$.

Can then $k < 0$, $k = 0$ and $k > 0$ be given physical meaning? The inequality $k < 0$ can tentatively be associated with radiation pressure countering gravity due to the cosmic background radiation; $k = 0$ with an absolutely void space-time; and $k > 0$ with homogeneously distributed cosmic dust or cosmic matter of unknown composition, reinforcing gravity.

References

1. A. Lightman and R. Brawer, *Origins: The Lives and Worlds of Modern Cosmologists* (Harvard University Press, Cambridge, MA, 1990).
2. R. A. Alpher and R. Herman, *Genesis of the Big Bang* (Oxford University Press, Oxford, 2001).
3. A. A. Penzias and R. W. Wilson, *Ap. J.* **142** (1965) 419.

Chapter 9

The Einstein–Lemaitre Correspondence

The cosmological expansion is, because its scale, the most grandiose phenomenon in nature. As we have seen, it was predicted theoretically[1] by the Russian mathematicians Alexander Alexandrovich Friedmann (1888–1925) in 1921 and, independently, by the Belgian astrophysicist and Catholic priest Georges Edouard Lemaitre (1894–1966) in 1927.

In October 1927, at Brussels, during the Solvay Conference on Physics (October 24–29) Albert Einstein met a relatively young Lemaitre (then 33), a Catholic priest wearing Roman collar, who ten years before had been an artillery officer in the Belgium Army at World War One. At the time of their encounter, Einstein probably did remember Lemaitre's paper[2] on solutions of his relativistic cosmological equations. This paper may have made him think of the article by Friedmann, the Russian mathematician who had obtained solutions for his cosmological equation which implied an expanding universe.

Einstein was then a little rude with the young cosmologist and Catholic priest: "Your calculations are correct, but your physics is abominable."[3] This is the same reaction he had to the work of the

Russian mathematician in 1922 (Friedmann had died in 1925; his disciple George Gamow, who would develop twenty years later Friedmann's model of the expanding universe, popularized the Big Bang model).

By 1927 Einstein was beginning to consider seriously the mathematical correctness of Friedmann and Lemaitre's expanding solutions of his cosmological equations, but he was still unconvinced. By then, however Lemaitre had gone much further than Friedmann. After spending a year in Cambridge, UK, with Arthur S. Eddington, the first champion of Einstein's theory of relativity in England (who became a world celebrity after the African eclipse expedition which detected the bending of starlight by the Sun), Lemaitre was encouraged by Eddington to travel to Cambridge, Massachusetts, for another year, to work with Harlow Shapley, who had been mapping the Milky Way using the 60-inch telescope of Mount Wilson, California.

There were already strong indications that cosmic nebulae far away were receding from us in space. This was taken at least as preliminary experimental evidence of an expanding cosmos. As noted by Farrel,[3] Lemaitre accompanied Einstein in Brussels to Piccard's lab and brought up the subject of recent measurements by American astronomers on spiral nebulae, suggesting that the redshifts in their spectra were indicative of a general expansion of galaxies in the universe. Two years later Einstein did begin to consider the possibility that the overall expansion of the universe was real. Within about two years, he had begun to accept the expansion publicly.

According to Helge Kragh[4]

In his (Lemaitre's) 1933 address (to the American National Academy of Sciences) he suggested an interesting interpretation of the (cosmological) constant, namely, that it may be understood as a negative vacuum density: "Everything happens as though the energy in vacuo would be different from zero", he wrote, referring to general relativity applied to regions of space of extremely low density. He then argued that in order to avoid a nonrelativistic vacuum or ether — a medium in which absolute motion is detectable — a negative pressure $p = -\rho c^2$ must be introduced, the vacuum density being related to the cosmological constant by $\rho = \Lambda c^2/4\pi G$. This negative *pressure is responsible for the exponential (de Sitter) expansion of Lemaitre's universe during its last phase, and it contributes a repulsive cosmic force of $\Lambda c^2 r/3$.*

(Here ρ = density, c = speed of light, Λ = cosmological constant, G = Newton's constant of gravity).

Note that, as pointed out in the previous chapter the *negative vacuum density* of Lemaitre can be advantageously substituted by *the radiation pressure due to the CBR* (cosmic background radiation) associated either to the negative space-time curvature (k < 0), to the cosmological constant ($\Lambda \neq 0$) or to both.

After their first meeting in Brussels in 1927, Lemaitre met Einstein again several times and corresponded for many years afterwards. Einstein's correspondences with Lemaitre on the famous cosmological constant are given in the first page of the letter which he[5] wrote to Lemaitre in September 28, 1947:

Dear Professor Lemaitre,

I thank you very much for your kind letter of July 30. In the meantime I received from Professor Schlipp your interesting paper for his book. I doubt that anybody has so carefully studied the cosmological implications of the theory of relativity as you have. I can also understand that in the shortness of T_o (present "age" of the universe, NT) there exists a reason to try bold extrapolations and hypothesis to avoid contradiction with facts. It is true that the introduction of the Λ term offers a possibility, it may even be that it is the right one.

Since I have introduced this term I had always a bad conscience. But at that time I should see no other possibility to deal with the fact of the existence of a finite mean density of matter. I found it very ugly indeed that the field law of gravitation should be composed of two logically independent terms which are connected by addition. About the justification of such feelings concerning logical simplicity it is difficult to argue. I cannot help to feel it strongly and I am unable to believe that such an ugly thing should be realized in nature.

In his book *The Primeval Atom*,[6] Lemaitre says:

We can compare space-time to an open, conic cup... The bottom of the cup is the origin of atomic disintegration (the disintegration of the "primeval atom", NT); it is the first instant at the bottom of space-time, the now which has no yesterday (*emphasis added*) *because, yesterday, there was no space.*

In my opinion, Einstein was right in dismissing his cosmological constant, but, given the effective equivalence between the *space-time curvature term with* $\Lambda > 0$, Lemaitre was perhaps not completely wrong.

References

1. M. Heller and A. Chernin, *Los Origenes de la Cosmologia: Friedman y Lemaitre,* Translated to Spanish from the Russian edition (Editorial URSS-Libros de ciencia, Moscú, 1991).
2. G. Lemaitre, *Ann. Soc. Sci. Brux. A* **47** (1927) 49, p. 12.
3. J. Farrell, *The Day without Yesterday* (Tunder's Mouth Press: New York, 2005), p. 10.
4. H. Kragh, *Cosmology and Controversy: The Historical Development of Two Theories of the Universe* (Princeton University Press, Princeton, N.J., 1996), p. 53.
5. Einstein's letter to Lemaitre in September 1947 (Albert Einstein Archives, Jewish National and University Library); reproduced in Ref. 4, p. 143.
6. G. Lemaitre, *The Primeval Atom: A Hypothesis of the Origin of the Universe,* Translated by B. H. Korff and S. A. Korff (D. Van Nostrand, New York, 1950).

Part III: More Paradoxes

10. The Accelerating Universe Paradox
11. The Missing Mass and Dark Energy Paradoxes
12. The Photon-to-Baryon Ratio Paradox

My problem is solving equations but what is the meaning of the solutions... let others think about it.

Alexander Friedmann
(1888–1925)

The world is finite. Otherwise we would be obliged to affirm something which, in itself, is outside our possibility of knowledge.

Georges Lemaitre
(1894–1966)

Chapter 10

The Accelerating Universe Paradox

The paradox of the accelerated expansion of the universe can be stated as follows: if, after the Big Bang, expanding matter (mainly galaxies but perhaps all other kinds of mass matter) is moving radially away from the center of gravity of the finite mass universe, in the absence of any increasing counter-gravitational force, why are not galaxies seen as slowing down under the powerful central gravitational attraction?

The CBR, which at earlier stages provided sufficient radiation pressure to drive the expansion, is cooling down with time. So it could not be causing cosmic acceleration. That is why, when accelerated expansion was reported in 1998, the specter of the cosmological constant, which had already undergone several ups and downs in popularity, was immediately called upon to justify the observations of the Type Ia supernovae suggesting the accelerated expansion. In the following years several observations corroborated this suggestion.

Before analyzing the Type Ia supernovae observations, it is necessary to specify clearly where we take the origin of space-time. The origin of space-time should be taken at the earliest cosmic sphere lying beyond the sphere of the CBR. At the time of decoupling the primordial plasma universe became a universe of atoms and the first light (the CBR) began to propagate within the growing cosmic sphere. Our galaxy could be considered to be relatively near the origin of the spherical space occupied by the background radiation. The light emitted long ago by the supernovae in relatively far away galaxies (therefore nearer the space-time origin) suggests that the rate of change of redshift with distance was slower at $z \simeq 0.5$ than it is now at ($z \simeq 0$). What is the role to be expected for the finite speed of light in connection with this behavior? Let us see.

Information is coming out to our planet from galaxies with small redshifts (say $z = 0.01$), from farther away galaxies (with redshifts $z \simeq 0.5$), and from even more distant galaxies or quasars (up to $z \simeq 1$ or more). This reflects indirectly the time-evolution of the cosmic expansion. In order to analyze cosmic dynamics, it is necessary always to keep in mind the cosmic evolution of the main cosmic parameters, $\Omega(t) = \rho(t)/\rho_c(t)$ and $H(t) \cdot t = (\dot{R}/R) \cdot t$, where t is the time elapsed since the Big Bang, taking into account that $\dot{r}/r$ for any galaxy is not necessarily the same as $\dot{R}/R$ for the radius of the universe.

At $t < t_0$ (for instance $t = 10, 3$ Gyrs) the distance r from our galaxy to other galaxies, and the recession velocity $\dot{r}/c$ from them, would be

$$0 \leq r \leq R, \quad 0 \leq \dot{r}/c \leq 1.$$

At $t = t_0$ (i.e., now, $t_0 = 13.7 \, \text{Gyrs}$) the respective distance and velocity are

$$0 \leq r_0 \leq R_0, \quad 0 \leq \dot{r}_0/c \leq 1.$$

And at a time $t > t_0$ in the future (for instance $t = 15 \, \text{Gyrs}$) the respective distance and velocity should be

$$0 \leq r \leq R, \quad 0 \leq \dot{r}/c \leq 1.$$

At present r_0 is the distance to the supernova when the light arriving *now* to us was emitted, and the velocity $\dot{r}_0 = v_0$ (deduced from the redshift) is the recession velocity when the light arriving to us *now* was emitted. R_0 and $\dot{R}_0$ on the other hand are the cosmic radius (*now*) and its growth velocity (*now*).

Let us analyze supernovae data[8] ($1\text{pc} = 3.07 \times 10^8$ cm) taking into account that $m[r(\text{pc})]$, the apparent magnitude, is given by

$$m = 5 \log_{10} r(pc) + (M - 5) \tag{10.1}$$

where $M = M_{\text{sn}} = -19.3$ is the relevant absolute magnitude to obtain the distance $r(\text{pc})$. On the hand, knowing z (redshift) is sufficient to obtain

$$\frac{\dot{r}}{c} = \frac{(z+1)^2 - 1}{(z+1)^2 + 1} \leq 1 \tag{10.2}$$

which gives the recession velocity divided by the speed of light. The Hubble ratio $H = \dot{R}/R$ does not necessarily coincide, as previously noted, with the ratio $h = \dot{r}/r$, except for galaxies in our close neighborhood ($z \ll 1$). For sufficiently distant supernovae, $z \approx 1$ or more, because of the relativistic relationship between ($\dot{r}/c$) and z given above, which is a consequence of the finite value of the velocity of light, we expect that $h < H$.

Figure 10.1 represents in log-log form a set of velocity versus distance data[8] for a number of supernovae with redshifts in the interval z from 0.01 to 1. For $z \ll 1$ the velocity versus distance relationship should be given by Hubble's law $H = \dot{R}/R$ see Eq. (7.12),[1] which is specified in Fig. 10.1 as $H = 87.75$ km/s/Mpc for $t = 10.3$ Gyrs, $H_o = 67.9$ km/s/Mpc for $t_o = 13$ Gyrs (present), and $H = 61.4$ km/s/Mpc for $t = 15$ Gyrs. In all cases, we have represented $\dot{r}$ versus r taking into account the relativistic expression for $v/c \equiv \dot{r}/c$ given in Eq. (10.2).

The data points corresponding to the supernovae measured in 1998 up to $z = 1$ are well fitted by Hubble's law ($\dot{r}/r = $ constant) up to $z \approx 0.1$, but begin to deviate thereafter, confirming the expectation that (v/c) must begin to saturate towards (v/c) ≈ 1 at some distance.

Let us see how relativistic considerations can be used to get the recession velocity in terms of $\dot{R}_o(0)$ and $\dot{R}(r_o)$, where r_o is the distance to the supernovae when the light arriving to us now was emitted. According to the Special Theory of Relativity

$$v = \frac{\dot{R}_o(0) - \dot{R}(r_o)}{1 - \dot{R}_o(0) \cdot \dot{R}(r_o)/c^2}. \tag{10.3}$$

Then, assuming that the change $d\dot{R}(r_o)$ is proportional to $-\dot{R}(r_o)dr_o$, we get

$$\dot{R}/\dot{R}_o \approx e^{-2r_o/r_m} \tag{10.4}$$

where r_m is a characteristic distance. Then

$$v \equiv v\left(\frac{r_o}{r_m}\right) = \dot{R} \cdot \left[\frac{1-\dot{R}/\dot{R}_o}{1-(\dot{R}_o^2/c^2)\cdot\dot{R}/\dot{R}_o}\right]$$

$$\tag{10.5}$$

$$= \dot{R}_o \cdot \left[\frac{1+\dot{R}/\dot{R}_o}{1-(\dot{R}_o^2/c^2)\cdot\dot{R}/\dot{R}_o}\right]\left(\frac{1-\dot{R}/\dot{R}_o}{1+\dot{R}/\dot{R}_o}\right) = \alpha\tanh(r_o/r_m)$$

where, for $R \to R_o$, $(r_o \ll R_o)$ i.e., $\dot{R} \to \dot{R}_o$, we get

$$\alpha_o \approx \dot{R}_o\frac{2}{1+\left(\dot{R}_o^2/c^2\right)} \approx H_o r_m \tag{10.6}$$

and for $R \ll R_o$, $(r_o$ comparable to $R_o)$, i.e., $\dot{R} \ll \dot{R}_o$, we get

$$\alpha_m \approx \dot{R}_o \tag{10.7}$$

where $r_m = R_o - R_m$ is the supernovae distance at which the recession velocity begins to saturate towards $v = v_{max}$. Thus, v_m must be somewhere below c, because galaxies take some time to form after the Schwarzschild radius R_+, at which $R_+ \approx c$. Then

$$\frac{v}{c} = \frac{v_m}{c}\tanh\left(\frac{r_o}{r_m}\right) \tag{10.8}$$

In Fig. 10.1 (v/c) as a function of r is given for three values of Hubble's parameter in all three cases as $(v_m/c)\tanh(\gamma r)$, where γ is the inverse of the characteristic distance r_m.

The data in Fig. 10.1 do not allow an accurate determination of the saturation velocity corresponding to the oldest observable galaxies, but they seem to indicate roughly a value $(v_m/c) \approx 0.72$ not far from one. Of course the general trend of the recession velocity of the galaxies, as shown in Fig. 10.1, implies that the rate of change of the observed galaxy velocity at distances of the order of $r \approx r_m$ is *smaller* than the rate of change of the observed galaxy velocity in our close neighborhood.

r(pc)

Fig. 10.1.   Apparent recession velocity, $\dot{r}/c = v/c$, versus apparent distance r(pc), fitted by $v/c = (v_m/c)\tanh(r/r_m) = (v_m/c)\tanh(\gamma r)$, where $\gamma \equiv 1/r_m$. Keeping only the nine points pertaining to the "High z Supernova Search" results in a much better fit with the above equation, which is not shown in Fig. 10.1 probably because α in Eq. (10.6) is somewhat larger than α in Eq. (10.7).

But it would be misleading to conclude from this that the cosmic radius $R(t)$ is really accelerating. In fact the value of the Hubble parameter, as shown in Fig. 10.1, decreases with time from $H = 87.75$ km/s/Mpc at $t = 10.3$ Gyrs, to $H_o = 67.9$ km/s/Mpc at $t_o = 13$ Gyrs (present), to $H = 61.4$ Km/s/Mpc at $t = 15$ Gyrs, as given by

$$H = \frac{\dot{R}}{R} = \frac{|k|^{1/2}c}{R_+} \cdot \frac{\cosh y}{\sinh^3 y}, \tag{10.9}$$

obtained by means of the compact Friedmann–Lemaitre solutions of Einstein's cosmological equation in Chapter 7.

In other words, the reported acceleration of galaxies in our close neighborhood, $z \approx 0$ ($r_o \approx R_o$), is not incompatible with a really decelerating growth of the cosmic radius, as given by the Friedmann–Lemaitre solutions, Eqs. (7.10) and (7.11), and with a real decrease with time of the Hubble's parameter, as predicted by Eq. (7.12).

References

1. A. Reiss *et al.*, *Astronomical Journal* **116** (1998) 1009.
2. S. Perlmutter *et al.*, *Astrophysical Journal* **517** (1999) 565.
3. A. G. Reiss *et al.*, *Astrophysical Journal* **607** (2004) 665.
4. D. N. Spergel *et al.*, *Astrophysical Journal Suppl.* **148** (2003) 175.
5. B. Chaboyer and L. M. Krauss, *Astrophysical Journal Letters* **567** (2002) L4.
6. W. M. Wood-Vasey *et al.*, *Astrophysical Journal* **666** (2007) 694.
7. P. Astier *et al.*, *Astronomy and Astrophysics* **447** (2006) 31.
8. S. Perlmutter, *Phys. Today*, April (2003) 56.

Chapter 11

The Missing Mass and Dark Energy Paradoxes

The *critical density* is the cosmic density (mass per unit volume) at which the speed of growth of the cosmic radius ($\dot{R}$) is exactly at escape velocity, $\rho_c = 3H^2/8\pi G$, where $H = \dot{R}/R$ is Hubble's parameter. At present, the *actual cosmic density* in our close neighborhood ρ_0, estimated on the basis of ordinary matter, luminous matter from galaxies and cosmic dust, is only about 4.5% of the critical density. Many theorists, specially inflationary theory supporters, are clearly in favor of a ratio $\rho_0/\rho_c = 1$ corresponding to a flat ($k = 0$) universe. This 100% of total equivalent mass would include as much as 25% *dark matter* (of unknown provenance) and about 70% *dark energy* (of even more unknown nature) associated with the cosmological constant.

Missing mass, or dark mass, was postulated by Fritz Zwicky[1] 75 years ago to account for the non-Keplerian velocities of orbital motion around the center of mass in galaxies and galaxy clusters. Later, gravitational lensing of background cosmic objects by galaxy clusters, and the temperature distribution of hot gas within galaxies and galaxy clusters were proposed as supporting large amounts of dark mass in the universe. It should be noted that most of these reports did not consider possible cosmic density variations with distance (with increasing densities at earlier times, further distances) for sufficiently old and distant early galaxies.

After the advent of inflationary cosmology and the observational reports suggesting considerable amounts of dark matter, many cosmologists accepted the hypothesis that most matter in the universe is non-baryonic, i.e., not made up of protons and neutrons within nuclei. This mysterious dark matter was supposed not to interact with ordinary matter via electromagnetic or nuclear (strong or weak) interactions. The possibility of "hot" dark matter, "warm" dark matter, and "cold" dark matter, or a combination of the three kinds of dark matter, have been considered[2] in speculative discussions of the problem during the last four decades.

The search for tangible dark matter, however, has not been very successful.[3] According to J. Silk, cold dark matter has been seriously challenged. Some amount of non-baryonic dark matter might be homogeneously distributed throughout cosmic space, but to conclude from the observations that its amount is of the order of 26%, and that it is made up of elusive, non-tangible hypothetical particles or objects, is not very realistic. The list of potential candidates during the last three or four decades includes massive neutrinos (now more or less discarded), WIMPS (weakly interactive massive particles), MACHOS (massive compact halo objects), and massive neutron star binaries, considered to be responsible for abundant GBR (gamma-ray burst) sources in the universe. If these or other possible candidates are as abundant in the universe as to account for one quarter of its mass they should have been detected and identified long ago. It may be also pointed out that M. Milgrom[4] has proposed a modification of Newtonian dynamics as a possible alternative to the "missing" mass hypothesis. According to R. A. Alpher and R. Herman[5] it is undisputable that, at early times, the density parameter $\Omega = \rho/\rho_c$ for an open universe with $\Lambda = 0$ should be very near 1. In their words:

> *We have evaluated Ω numerically and find, as expected, that at early times the value of Ω is extraordinarily close to unity and that the deviation from unity becomes increasingly small for earlier and earlier times.*

As we have seen in Chapter 7, from the compact Friedman–Lemaitre solutions [Eqs. (7.8) and (7.9)] for and open universe ($k < 0$, or equivalently, for $\Lambda = -3k^2/R < 0$), at very early times Ω (dimensionless) approaches 1, and $H \cdot t$ (dimensionless) approaches 2/3.

Alpher and Herman conclude that:

> *It may not be necessary to invoke that consequence of an inflationary paradigm which requires that the value of Ω be unity throughout the history of the expansion ...*

Dark energy,[6] necessary to complement dark matter in order to get $\Omega = 1$, is assumed to be a certain kind of potential energy permeating all space-time and tending to increase the expansion rate at later times with respect to earlier times, the times at which the first protogalaxies were formed. It is usually assumed to be associated[7] with a nonvanishing "cosmological constant" and is sometimes connected with hypothetical scalar fields (quintessence). The cosmological constant is often tentatively explained as physically due to the vacuum energy (the zero-point energy of the vacuum). This is not so easy to do without incurring notable numerical inconsistencies in its order of magnitude. Measuring the equation of state of this dark energy is currently one of the main tasks of observational cosmology, but for the moment, the effort has not been very successful.

Thus, the dark matter and dark energy that have not yet been observed, constituting about 95% of the universe matter/energy content, which have been intensely but unsuccessfully investigated for more than twenty years, can be said to constitute one of the greatest standing cosmic paradoxes at the present time. There is the danger, however, that after so many years, cosmologists may get so used to them that begin to think that they are not at all paradoxical, but quite natural.

Would it be possible, by taking into account the *time-evolution* of the mass matter density parameter ratio $\Omega_m(y) = \rho_m(y)/\rho_c(y)$ and the space-time curvature potential energy density parameter ratio $\Omega_k(y) = |k|(c/\dot{R}(y))^2/\rho_c(y)$, that the mysteries of the dark matter and the dark energy be somehow solved?

Equation (7.1) can be written as

$$\rho_c = \rho + \rho_c |k| \left(\frac{c}{\dot{R}(y)} \right)^2 \tag{11.1}$$

or equivalently, as

$$1 = (\Omega_m(y) + \Omega_\gamma(y)) + \Omega_k(y) = \Omega(y) + \Omega_k(y), \tag{11.2}$$

where, for $y > 0.881$ ($z > 20.7$, $T > 59.3\,\mathrm{K}$),

$$\Omega_m = \Omega \left(1 + \frac{T}{T_{\mathrm{af}}} \right) \approx \Omega = 1 - \tanh^2 y, \tag{11.3}$$

$$\Omega_k = 1 - \Omega = \tanh^2 y, \tag{11.4}$$

with $y \equiv \sinh^{-1}(T_+/T)^{1/2}$, $T_+ = 59.2\,\mathrm{K}$, $T_{\mathrm{af}} = 2968\,\mathrm{K}$.

Table 11.1.   The evolution of Ω_m (matter), Ω_r (radiation) and Ω_k (space-time curvature potential energy) as a function of time t (Gyrs) and temperature T(K).

y	(z)	t(Gyrs)	T(K)	Ω_m	Ω_k
2.243	0	13.70	2.726	0.044	0.956
2.196	0.100	12.36	3.000	0.048	0.951
2.057	0.467	9.08	4.000	0.063	0.936
1.949	0.834	7.14	5.000	0.077	0.922
1.862	1.201	5.83	6.000	0.091	0.908
1.789	1.567	4.92	7.000	0.105	0.895
1.726	1.934	4.23	8.000	0.118	0.881
1.671	2.301	3.70	9.000	0.131	0.869
1.400	4.990	1.85	16.33	0.215	0.783
1.200	8.53	1.05	26.00	0.302	0.695
1.100	11.18	0.775	33.22	0.355	0.640
0.881	**20.72**	**0.365**	**59.23**	**0.490**	**0.500**
0.800	26.54	0.266	75.09	0.545	0.440
0.700	36.75	0.173	102.92	0.613	0.365
0.600	52.60	0.106	146.20	0.678	0.288
0.400	127.80	0.030	251.23	0.765	0.144
0.140	1087.0	0.001	2968.00	0.990	0.019

Note that for $y = 0.140$, $T = T_{\mathrm{af}} = 2968.0\,\mathrm{K}$ (decoupling, atom formation), $\Omega_m = \Omega_\gamma = 0.490$, i.e., *mass, matter* and *radiation matter* energy densities are equal and are together near one, and $\Omega_k = 0.019$ is much smaller than one.

From the beginning of the galaxy formation time ($T = T_+ = 59.2\,\mathrm{K}$) to the present ($T = T_0 = 2.726\,\mathrm{K}$), Ω_m decreases from nearly $1/2$ to nearly 0.044, and Ω_k increases from nearly $1/2$ to nearly 0.956, which lead to the following average values

$$\langle \Omega_m \rangle \approx 0.272,$$

$$\langle \Omega_k \rangle \approx 0.728.$$

The figures are surprisingly close to the current estimates for *dark matter* and *dark energy*. If the argumentation is correct (and why not) a lot of research money could be saved.

Note that, regardless of the increase in local matter mass density to be expected as $r = R_0 - R$ approaches $r = R_0 - R_+$, the actual matter mass density at any time (in a homogeneous universe) is assumed to be the same everywhere, and the total mass of the universe must therefore be the same at $t = t_+ (R = R_+)$, when the first protogalaxies begin to form, and at $t = t_o (R = R_o)$, the present: with no "missing mass", properly speaking.

The source of the space-time curvature energy density, which is $|\mathrm{k}|$, is also *independent of time*, even if Ω_k increases with time at the same rate as $\Omega = \Omega_m + \Omega_r$ decreases.

The key is to take properly into account the time-evolution of $\Omega(\mathrm{t})$ and $H(\mathrm{t})$.

In June 15, 2001 a reanalysis of WMAP's data by astronomers in the Physics Department at Durham University suggested that the conventional wisdom (dark energy 74%, dark matter 22%) about the content of the universe may be wrong.

The report[8] "Cross-correlating WMAP5 with 1.5 million LRGs: a new test for ISW effect", co-authored by astronomers from Durham University, Anglo-Australian Observatory, University of Sydney, The Pennsylvania State University and Yale University, was published in the prestigious British Journal *Monthly Notices of the Royal Astronomical Society*.

If their findings are confirmed they may end up contributing to solve satisfactorily the "dark energy" and "dark matter" paradoxes.

References

1. F. Zwicky, *Helvetica Physica Acta* **6** (1933) 110–127.
2. J. Silk, *Physics Reports* **405** (2005) 279.
3. J. Silk, *Europhysics News* **32**(6) (2001) 211.
4. M. Milgrom, *Astrophysical Journal* **270** (1983) 365–370.
5. R. A. Alpher and R. Herman, *Meeting on Unified Symmetry in the Small and the Large* (Coral Gables, Florida, 1995).
6. P. J. E. Peebles and B. Ratra, *Reviews of Modern Physics,* **75** (2003) 559–606.
7. S. Carroll, *Living Reviews in Relativity,* **4** (2001) 1.
8. U. Sawangwit *et al.*, *Monthly Notices of the Royal Astronomical Society,* **402**(4) (2009) 2228–2244.

Chapter 12

The Photon-to-Baryon Ratio Paradox

In his book[1] *The First Three Minutes*, Steven Weinberg says:

> *We observed in the preceding chapter that there is now a nuclear particle for every 1000 million photons in the microwave radiation background (the exact figure is uncertain) so the baryon number per photon is about one billionth* (10^{-9}).

After noting that no one has seen any sign of appreciable amounts of antimatter in the universe, Weinberg considers the possibility that the density of photons has not remained proportional to the inverse cube of the size of the universe:

> *This could happen if there were some sort of departure from thermal equilibrium, some sort of friction or viscosity which could have heated the universe and produced extra photons. In this case, the baryon number per photon might have started at some reasonable number, perhaps around one, and then dropped to its present low value as more photons were produced.*

This could be called the *baryon-to-photon paradox* (10^{-9}) or the *photon-to-baryon paradox* (10^{9}).

At the time this was written the expansion and cooling down of the universe was generally assumed to be described by a *single equation of state*. However, this was not very realistic.[2]

There is a distinct change in the universe at T_{af} (atom formation), i.e., about $3\,000$ to $4\,000$ degrees K, from a *plasma universe* $(T > T_{af})$ to an *atom universe* $(T < T_{af})$. The former is *opaque* to radiation, therefore, in it charged nuclei (mainly H and ^{4}He nuclei) and electrons scatter photons violently (friction) and extra photons are continuously produced. On the other hand, the atomic universe (at which, in due time, stars and galaxies would form) is basically *transparent*. Then the conservation of the number of photons (as well as that of stable nuclei) is to be expected thereafter through the expansion.

This implies a change in *equation of state* from the plasma universe, in which radiation mass density and matter mass density expand in *unison* $(T > T_{af})$ as given by

$$\rho_r(R) = \rho_m(R), \text{ resulting in } RT^{4/3} = R_{af}\, T_{af}^{4/3} \qquad (12.1)$$

of the present "atomic" universe, in which the number of photons $n_r(R)$ and of baryons $n_b(R)$ are both conserved through the expansion $(T < T_{af})$ and are given by

$$n_r(R) = n_b(R), \text{ resulting in } RT = R_{af}\, T_{af}. \qquad (12.2)$$

This implies that the number of photons equals the number of baryons at the characteristic baryon temperature $T_b = m_b c^2 / 2.8\, k_B$, $T > T_{af}$ (plasma universe) and we have:

$$\frac{n_r}{n_b} = \frac{\sigma T^4 / 2.8\, k_B T}{[M/\frac{4\pi}{3}R^3]/m_b} = \left(\frac{\sigma T^4}{[M/\frac{4\pi}{3}R^3]\cdot c^2}\right)\frac{m_b c^2/2.8\, k_B}{T} = \frac{T_b}{T}$$

$$(12.3)$$

since $\rho_r = \rho_m$ during the plasma phase of the universe $(T > T_{af})$. Therefore, at $T \approx T_b$ (baryon equilibrium temperature) we have

$$\frac{n_r(T_b)}{n_b(T_b)} \approx 1 \qquad (12.4)$$

as might be expected.

The cosmic *neutron-to-proton* ratio determined from the *relative abundance* of ^{4}He in the universe is

$$(n/p)_{\text{obs}} \approx 0.13 \pm 0.02. \tag{12.5}$$

If we assume that it is due to the spontaneous disintegration of primordial neutrons ($T_n = m_n c^2/2.8\,k_B \approx 3.88 \times 10^{12}\,\text{K}$) into protons, electrons and neutrinos once cosmic conditions are apt to allow free electrons ($T_e = m_e c^2/2.8\,k_B \approx 2.11 \times 10^9\,\text{K}$) we can use the known half-life of the neutron $\tau = 887\,\text{s}$ to calculate the neutron-to-proton ratio by means of

$$(n/p) \approx e^{-(t_{\text{ns}} - t_e)/\tau} \tag{12.6}$$

where t_{ns} is the cosmic time at the end of nucleosynthesis ($T_{\text{ns}} \approx 0.5 \times 10^9\,\text{K}$), with an estimated error of $\pm 20\%$, t_e is the cosmic time at the electron equilibrium temperature, and τ the free neutron half-life.

At this early epoch ($y \ll 1$) cosmic time is given by

$$t(y) \approx \frac{R_+}{|k|^{1/2}c} y^3 \tag{12.7}$$

which, taking into account that

$$y \equiv \sinh^{-1}\left[\frac{R}{R_+}\right]^{1/2} = \sinh^{-1}\left[\left(\frac{T_+}{T_{\text{af}}}\right)\left(\frac{T_{\text{af}}}{T}\right)^{4/3}\right]^{1/2}$$

$$\approx \left(\frac{T_+}{T_{\text{af}}}\right)^{1/2}\left(\frac{T_{\text{af}}}{T}\right)^{2/3} \tag{12.8}$$

reduces to

$$t = \left\{\frac{R_+}{|k|^{1/2}c}\left(\frac{T_+}{T_{\text{af}}}\right)^{3/2}\right\}\left(\frac{T_{\text{af}}}{T}\right)^2 = 6.09 \times 10^{13}\left(\frac{2968}{T}\right)^2 \tag{12.9}$$

where $R_+ = 4.588 \times 10^{26}\,\text{cm}$, $|k^{1/2}| = 0.7071$, $c = 3 \times 10^{10}\,\text{cm/s}$, $T_+ = 59.2\,\text{K}$ and $T_{\text{af}} = 2968$.[3]

The cosmic times at neutron formation (t_n), electron formation (t_e) and at the end of nucleosynthesis (t_{ns}) would then be given by

$$t_n = 3.56 \times 10^{-5}\,\text{s} \qquad (12.10)$$

$$t_e = 120.5\,\text{s} \qquad (12.11)$$

$$t_{ns} = 2145 \pm 400\,\text{s}. \qquad (12.12)$$

Therefore

$$(n/p)_{\text{cal}} \approx e^{-(2145\pm400-120.5)/887} = 0.10 \pm 0.04, \qquad (12.13)$$

which agrees reasonably well with the observed ratio given in Eq. (12.5).

We can check also that with Eq. (12.1), the cosmic equation of state for a *plasma universe*, the ratio of cosmic radii for closely packed electrons at T_e and closely packed baryons at T_b

$$\frac{R_e}{R_b} = \left(\frac{T_b}{T_e}\right)^{4/3} = \left(\frac{m_b}{m_e}\right)^{4/3} \approx 2 \times 10^4, \qquad (12.14)$$

is approximately the same as the ratio of particle radii. (For the electron radius taking into account Pauli's principle, we estimate $r_e \approx (1/2) \cdot r_{\text{Bohr}} \approx (1/2) \cdot 0.53 \times 10^{-8}$ cm. For the baryon radius we take $r_b \approx 1.24 \times 10^{-13}$).[4]

$$\frac{r_e}{r_b} \approx \frac{0.26 \times 10^{-8}}{1.24 \times 10^{-13}} \approx 2 \times 10^4, \qquad (12.15)$$

which is not the case using Eq. (12.2), the cosmic equation of state for a *transparent universe*, which would result in

$$\frac{R_e}{R_b} = \left(\frac{T_b}{T_e}\right) = \left(\frac{m_b}{m_e}\right) = 1836 \ll \frac{r_e}{r_b} \approx 2 \times 10^4. \qquad (12.16)$$

During the transparent phase of the expansion, radiation density (ρ_r) decreases much faster than matter density (ρ_m). In fact, at present $\rho_{mo}/\rho_{ro} \approx (T_{af}/T_0) \approx 1088$ within the same spherical space-time. Where is the missing radiation energy going? The answer may be in the parallel increase in potential energy, in particular the space-time curvature potential energy, at the expenses of radiation energy density, after the universe became transparent. At least this may be a tentative explanation.

References

1. S. Weinberg, *The First Three Minutes* (Bantam Books, New York, 1977), pp. 88–89.
2. N. Cereceda *et al.*, Is it realistic assume the same cosmic equation of state prior to and after atom formation? Frontiers of Fundamental Physics, Universidad Politécnica, Madrid 2006, *AIP Conference Proceedings*, **905**.
3. WMAP Collaboration, *Astrophys. J. Suppl.* **148** (2003) 13–15, 17, 18, 20, 22–25.
4. R. D. Evans, *The Atomic Nucleus* (Mc Graw-Hill, New York, 1955).

Part IV: A Contingent Universe

13. The Universe is Finite, Open and Contingent
14. The Very Early Universe: Indeterminacy or Uncertainty
15. Singular Moments in Cosmic History
16. A Brief Outline: World Events and Cosmological Discoveries from −4500 to 2010

In a world that was not the expression of intelligence science could never have come into being.

Edmund Whittacker
(1873–1956)

Chapter 13

The Universe is Finite, Open and Contingent

The universe is *finite*: As we have seen, it has a finite *mass*

$$M = 1.54 \times 10^{54} \, \text{g}, \tag{13.1}$$

a finite (but growing) *age* since the Big Bang,

$$t_o = 13.7 \times 10^9 \, \text{yrs}, \tag{13.2}$$

and a finite (but growing) *radius*, since that event,

$$R_o = 9.96 \times 10^{27} \, \text{cm}. \tag{13.3}$$

The finite mass $M = 1.54 \times 10^{54}$ g is conserved, of course through the cosmic expansion, while the actual time and the actual radius are finite, but increasing and, in an *open* universe, unbounded. We may note that the characteristic radius R_+, for a universe with a total matter mass given by $M = 1.54 \times 10^{54}$ g, with $y_+ = \sinh^{-1}, ^1$ is

$$t(y_+) = \frac{R_+}{|k|^{1/2} c} \{\sinh y_+ \cosh y_+ - y_+\} = 0.365 \times 10^9 \, \text{yrs}, \tag{13.4}$$

$$R(y_+) = R_+ \sinh^2 y_+ = 4.58 \times 10^{26} \, \text{cm}. \tag{13.5}$$

A *Compton* radius

$$r_c = \frac{\hbar}{mc}, \tag{13.6}$$

and a *Schwarzschild* radius

$$r_s = \frac{Gm}{c^2}, \tag{13.7}$$

can be associated to the universe as well as to various finite objects in it, going from finite galaxies to finite elementary particles.

The following Table gives *Compton* radius and *Schwarzschild* radius for various massive objects in the universe.

As we saw in Chapter 7, Eq. (7.16), if the Friedmann–Lemaitre solutions of Einstein's cosmological equations describe correctly cosmic evolution (and they do describe well the thermal history of the universe from the Big Bang to the present), the dimensionless product of the Hubble constant time the present age of the universe is given by

$$H_o t_o = 0.942 \pm 0.065, \tag{13.8}$$

which is incompatible either with a flat universe ($k = 0$), which requires

$$H_o t_o = 2/3, \tag{13.9}$$

or with a closed universe ($k > 0$), which requires

$$H_o t_o < 2/3. \tag{13.10}$$

The universe is therefore *open*, finite and unbounded. Modern physics tells us that we live in an evolving, finite and open universe.

Why is the universe what it is according to modern cosmology, and not anything else?

Table 13.1.   Compton and Schwarzschild radii for massive objects

Object	m(g)	r_c(cm)	r_s(cm)	r_s/r_c
Universe	$1.54 \cdot 10^{54}$	$2.26 \cdot 10^{-92}$	$1.14 \cdot 10^{26}$	$1.98 \cdot 10^{-118}$
Galaxy	$\sim 1 \cdot 10^{43}$	$3.5 \cdot 10^{-81}$	$7.41 \cdot 10^{-14}$	$0.47 \cdot 10^{-95}$
Star	$\sim 1 \cdot 10^{32}$	$3.5 \cdot 10^{-70}$	$7.41 \cdot 10^{3}$	$0.47 \cdot 10^{-73}$
Earth	$5.95 \cdot 10^{24}$	$5.85 \cdot 10^{-63}$	$4.43 \cdot 10^{-4}$	$1.32 \cdot 10^{-67}$
Planck monopole	$2.17 \cdot 10^{-5}$	$1.61 \cdot 10^{-33}$	$1.61 \cdot 10^{-33}$	1
Baryon	$1.67 \cdot 10^{-24}$	$2.09 \cdot 10^{-14}$	$1.23 \cdot 10^{-52}$	$1.23 \cdot 10^{38}$
Electron	$9.10 \cdot 10^{-28}$	$3.84 \cdot 10^{-11}$	$6.74 \cdot 10^{-56}$	$0.56 \cdot 10^{45}$

No purely physical theory and no purely physical experiment can give a concrete answer.

In other words the universe is *contingent*, it is not *necessarily* what it is, but it is *really* what it is, and not anything else. The term "contingent" implies a *physical* reality which cannot be measured directly, and also a *metaphysical* reality, more intangible but no less real, and which can be intellectually recognized.

And a contingent universe is a created universe

By the middle of the nineteenth century,[1] both the Hegelian *left* (Marx and Engels) and the Hegelian *right* (specially the neo-Kantians) had for a fundamental tenet that the universe (material or not) was *infinite*. Only a few first-rate scientists dared to disagree. Among them is *Gauss*, the prince of mathematicians, who noted that *Kant*'s dicta on categories were sheer triviality, probably keeping in mind non-Euclidean geometries. The finiteness of matter in endless space was implicit also in the work of *Riemann* and Zöllner.

Kant's *claim*[1] that the universe was a bastard product of the metaphysical cravings of the human intellect (put forward to discredit the classic cosmological argument to prove God's existence) was flatly denied with words and deeds by Planck and Einstein, the two greatest physicists of the twentieth century.

Planck, after liberating himself of Mach's tutelage, when he affirmed unambiguously[2] his full confidence in the reality of a *causally connected universe*.

Einstein, no less after finally emancipating himself from Mach's influence, produced the first contradiction-free treatment[1] of the totality of all gravitationally interacting objects which explicitly required a *universe with a finite mass*.

When Planck's son, Erwin, was executed for plotting against Hitler at the end of World War Two, everything seemed to have fallen in ruins around him: home, country, science. He wrote to a friend[3]:

> *What helps me is that I consider it a favor of heaven that since childhood a faith is planted deep in my innermost being, a faith in the Almighty and the All-good not to be shattered by anything. Of*

course his ways are not our ways, but trust in him helps us through the darkest trials.

In 1952, few years before his death, Einstein wrote to his friend Maurice Solovine[4]:

. . . that I think of the comprehensibility of the world. . . as a miracle (emphasis added) or an eternal mystery. But surely, a priori, one should expect the world to be chaotic. One might. . . expect that the world evidenced itself as lawful only so far as we grasp it in an orderly fashion. On the other hand, the kind of order created, for example, by Newton's gravitational theory is of a very different character. . . Therein lies the "miracle" which becomes more and more evident as our knowledge develops. . . And here is the weak point of positivists and professional atheists, who feel happy because they think that they have preempted not only the world of the divine but also of the miraculous. . .

As Stanley L. Jaki, perhaps the most incisive historian of Science of the twentieth century, writes[5] in his book *The Roads of Science and the Ways to God*, Planck and Einstein, with their confidence in the reality of a causally connected and finite universe, provide extraordinary compelling evidence in favor of a realistic metaphysics and epistemology, midway between idealism and positivism.

References

1. S. L. Jaki, *God and the Cosmologist* (Real View Books, Fraser, Michigan, 1998), pp. 13, 14, 17.
2. M. Planck, *A Survey of Physics*, translated by R. Jones and D. H. Williams (Methuen, London, 1925), pp. 1–41.
3. A. Herman, *Max Planck in Selbstzeugnissen und Bilddokument*, p. 98.
4. A. Einstein, *Lettres a Maurice Solovine* (Gauthier Villars, Paris, 1956), p. 102.
5. S. L. Jaki, *The Road of Science and the Ways to God* (University of Chicago Press, 1979), pp. 165–196.

Chapter 14

The Very Early Universe: Indeterminacy or Uncertainty

Quantum Mechanics, invented by de Broglie, Schrödinger, Heisenberg, Born, among others, to describe the physics of atoms and nuclei, is a statistical branch of physics,[1] and, as such, not very apt to describe the physics of the universe, a *singular entity*, absolutely unique, in spite of the current efforts to disguise it in a "multitude", called multiverse.

Heisenberg's principle,[2] which in 1927 specified correctly the inherent limitations to the accuracy of *complementary measurements* in the microscopic world of atoms and elementary particles (for instance position and momentum, time and energy), is based on principle's validity upon the specific value of Planck's constant, h or $\hbar = 1.05 \times 10^{-27}$ erg sec (the indivisible "quantum" of action). No doubt Heisenberg's principle has been extremely useful in describing the physical behavior of atoms, nuclei and particles; it can provide also unexpected and very significant insights about macroscopic physical systems; but has been also much abused by physicists, including Bohr and Heisenberg himself. Already in the very paper introducing his principle Heisenberg concluded that "the invalidity or at least the meaninglessness of the law of causality seems to be firmly established through recent developments in atomic physics". He said, further, that the inability of physicists to measure nature *exactly* implied the inability of events to take place in nature *exactly*. As noted by Jaki,[2] here the same word "exactly" is taken in two very different meanings, one operational (observational) and one ontological (factual, independent of the observer). The first does not exclude the second. As it is well known, Einstein, and with him many other great twentieth century physicists, including de Broglie and Schrödinger, did not agree with Heisenberg in the arbitrary invalidation of the law of *causality* in nature. (Where could we, poor physicists, go in the investigation of nature without a reasonable *confidence* — Planck's word — in the reality of a *causally connected universe?*).

Very conspicuous cosmologists, like Stephen Hawking, and many others, have tried by all means to suppress a Creator and an origin in time for the universe. One of the tricks is using surreptitiously Heisenberg's principle to postulate a virtual fluctuation of energy for an infinitesimal period of time, and jump from there to an actual physical universe existentially stable for billions of years.

Sometimes Heisenberg's principle is called correctly as the *uncertainty principle*. After all, the only means we have at our disposal to measure electrons and photons are photons and electrons. Limitations in the ultimate accuracy of the measurements are to be expected. Sometimes, however, Heisenberg's principle is called, improperly, the *indeterminacy* principle, implying that physical reality is undecided, not knowing for sure where *to be* or where *to go* in space-time at any moment. This is not serious. Applying this way of reasoning to a microscopic event, such as the disintegration of an alpha particle, could be taken as a sleight-of-hand or a joke. But applying it to the universe is really too much.

Bohr, who was Heisenberg's mentor, did not see any problem when he said "one speaks of a choice on the part of nature".[3] No wonder that he choose the *yin* and *yang* as his coat of arms, demonstrating that the Copenhagen interpretation of Quantum Mechanics (philosophy, not physics) implied a shift from Western to Eastern cultural references. He held also that a great truth is a statement whose contradiction is also a great truth. This sounds enlightened but in reality it is not.

Regarding the singularity implied by Einstein's cosmological equation with respect to *time*, as it was noted in Chapter 4, Heisenberg's uncertainty principle provides a limit to how close one is allowed to approach the origin. It is

$$t_H \equiv \frac{\hbar}{M_u c^2} \approx 7.54 \times 10^{-103} \text{ s},$$

determined by the finite mass of the universe $M_u = 1.54 \times 10^{54}$ g and the values of the universal constants $\hbar$ and c. This is very small in comparison with Planck's time

$$t_{Pl} \equiv (\hbar G/c^5)^{1/2} = 5.37 \times 10^{-44} \text{ s}.$$

No cosmic clock will ever be available to measure time at t_h, but it does help to know that there is an earliest limit.

References

1. G. t'Hooft, Plenary lecture at the International Symposium of "Frontiers of Fundamental Physics", Universidad Politecnica, Madrid, 2006, *AIP Conf. Proc.* **905** (2007), pp. 84–102.
2. S. L. Jaki, *God and the Cosmologists* (Real View Books, Fraser, Michigan, 1998).
3. S. L. Jaki, *The Ways to God and the Road of Science* (Chicago University Press, Chicago, 1987), p. 203.

Chapter 15

Singular Moments in Cosmic History

The Big Bang

A few microseconds after the beginning of space-time, about 13.7 billion years ago, baryons (neutrons) and high energy photons, at a temperature of $T \sim 4 \times 10^{12}$ K, were already expanding rapidly. A few minutes after, electrons were formed, and, in about 30 minutes, nucleosynthesis (mainly of ^{4}He nuclei) took place. The primordial ratio of neutrons-to-protons in the universe was fixed.

A few million years after, the expanding universe had cooled enough, to $T \sim 3 \times 10^3$ K, and atoms (mostly hydrogen but also substantial amounts ^{4}He) were formed. Matter and radiation decoupled and the universe became transparent.

After about 360 million years, stars and galaxies began to form.

The Sun and the solar planetary system

About five billion years ago, i.e., some nine billions years after the Big Bang, the Sun, a second or third generation star in the Milky Way, was formed out of cosmic dust from previous stars, which contained already substantial amounts of heavy elements from previous supernovae explosions. The solar system, made up by Mercury,

Venus, the Earth, Mars, Jupiter, Saturn, Uranus and Neptune, plus the asteroids, plus the comets and meteorites, was formed.

The Earth, located between Venus and Mars, occupied a privileged position, with a mass, a specific chemical composition and a magnetic field, which, in time, would allow our planet to develop a biosphere with plenty of water, and a surrounding gaseous atmosphere made up of oxygen and nitrogen.

The Earth–Moon system

The formation of planet Earth is dated to about 4 600 million years ago, and the formation of the Earth–Moon system, a lucky coincidence in all respects, is assumed to be the result of a cosmic clash between a very large foreign body and the primitive Earth. The Earth–Moon system is absolutely unique in our solar system. No other satellite has a mass comparable to that of the corresponding planet. This has important consequences: it stabilizes the rotation axis of the Earth and makes possible the four seasons during the year. It makes possible in our planet a night luminary for the most of the months. And, in due time, when the oceans and the Earth's atmosphere are formed, it makes the Moon the main agent of the tides, with important consequences.

The unique "habitat" made up by our *Earth–Moon system* depends on a number of physical *preconditions*.[1,2] In particular:

- The actual combination of specific universal *physical constants* which determines the scales of mass, size and time in the universe.
- The possibility of existence of stable and radioactive *nuclei* making up the Periodic Table of the elements.
- The possibility of existence of stable *atoms* and molecules at the temperatures on the Earth.
- The abundant availability of *light elements* (C, O, N, S) on the Earth's crust; heavy *non-metallic* elements (Mg, Ca, Str, Ba) in the mantle; and heavy *metallic* elements (Fe, Ni, Co) in the magnetic core, responsible among others things for the Earth's magnetic field.

- That the Earth is at the *right distance* from the Sun, such that its mean temperature is about 300 K, intermediate between the melting point of ice and the boiling point of water.
- That it has a *mass* and a *size* adequate to retain an atmosphere rich in O_2 and N_2 in the right proportion.

The list of physical and chemical preconditions could go on further. But this sample is enough to illustrate the differences with our two neighboring planets in the solar system: Venus and Mars. Kant and other eighteenth century enlightened philosophers expected intelligent beings in both. Today we know, through NASA's research satellites, that both are quite inhospitable to life.

Life

About 3 600 million years ago, the first fossil traces of life were left out in old carbonaceous rocks. The most elementary unicellular organism, a bacteria, has already a tremendous complexity. All living beings, from the bacteria to the most developed mammals, have a genetic code which, through its DNA, preordains its development, and through its RNA and its proteins, controls its metabolism, its immunologic defences and its reproductive system. The basic components of the cells constituting unicellular organisms, plants, and animals, are the same. And their genome is made up of long sequences of the same nucleotides: adenine, guanine, cytosine and thymine, linked by sugar and phosphate groups.

Francis Crick,[3] an agnostic, co-discoverer of the structure of DNA, and winner of the Medicine and Physiology Nobel Prize in 1962, says:

> *An honest man, armed with all the knowledge available to us now, could only state that in some sense, the origin of life appears at the moment to be almost a miracle, so many are the conditions which would have had to have been satisfied to get it going.*

A representative sample of these conditions[4] is the following:

- *Plate tectonics* and volcanic activity contributing to increase the H_2O content in the biosphere.

- An early, prolonged, *rain of meteorites* on the Earth's surface, contributing to the same end.
- The subsequent accumulation of *water*, with its exceptional physical and chemical properties, in the world's oceans.
- The extraordinary *chemical properties of C* which result in the possibility of organic chemistry.
- The exceptional combination of biochemical properties in the *amino acids*, the *proteins* and the *DNA macromolecules*.
- The marvel of *photosynthesis*, through which green plants take advantage of the solar flux during the day to synthesize organic matter, using CO_2 and liberating O_2.
- The subtle *biochemistry* that controls the metabolisms, the immunological system the reproductive mechanisms of the vertebrates.
- The fact that the Earth has a *large planetary companion* such as Jupiter, orbiting the Sun in the same plane, and protecting our planet from most asteroids and meteorites.
- The existence of a protective belt, the *Van Allen belt*, held in place by the Earth's magnetic field in the upper ionosphere, which protects our planet from too much cosmic radiation.
- The fortunate *location* of the Sun in an outer region of the *Via Lactea* in which the background galactic radiation is relatively low. As pointed out in *The Privileged Planet*,[4] the Sun's location and the Earth's location within the galaxy are ideal for a cosmic observatory, resulting in the possibility of observation of solar eclipses, and among other things, the testing of General Relativity.

Man

About 60 thousand years ago the last great glaciations took place. Some 15 thousand years ago, *primitive men*, with creative artistic skills comparable to those of Dalí or Picasso, painted bisons, horses and hunters in the walls of Altamira and Lascaux. Only 10 thousand years ago the inhabitants of primitive cities developed agriculture and domestication of animals (cattle, horses, camels, sheep).

There took place, as well, the first fabrication of ceramics and clay vessels.

It is *undeniable* that men share with the higher mammals, specially the apes, a great many physical characteristics. It is well known that the genome of the *chimpanzee* (living on Earth for more than five million years according to our contemporary anthropologists), differs from that of *men* only in *one percent.*

It is even more *undeniable* that men are quite different from ordinary apes:

- Men are the only apes that *speak*, in many languages (Spanish, English, Russian, Chinese...); the only ones that *write*, in many alphabets (Cuneiform, Roman, Cyrillic, Arabic...); the only ones that make *symbols*, like flags, logos, national anthems...
- Men are the only apes that truly appreciate the beauty of *sunset* or the splendor of a *starry night*; the only ones that have invented folk songs; and games, from football to chess; the only apes who have held Olympics Games, in ancient Greece and in our own twentieth century.
- Men have been the only apes able to built such large constructions as the *Great Chinese Wall* (it can be viewed by the astronauts from outer space), the *Egyptians Pyramids*, or the great medieval *European Cathedrals.*
- The only apes which have written *symphonies* and painted *masterpieces* such as the Sistine Chapel.
- The only ones which have written *epic poems* such as, the *Iliad*, or *novels* such as *Don Quixote*, or *dramas* such as *Hamlet.*
- The only apes which have *historical records*, or *calendars*, or *ships* to explore the five continents.
- Men are the only apes that have developed *scientific learning*: not only *grammar*, *logic* and *rhetoric*; but also *mathematics*: geometry and algebra; the only apes which, within the last three centuries, have developed *physics*: mechanics, electromagnetism, thermodynamics, and optics.
- Men are the only apes that have developed *advanced technologies*; can communicate at the speed of light; can travel in jets at

supersonic speeds; or have invented powerful computers, lasers and GPS.

- Men are the only apes — for the time being — which have been able to *fly to the Moon.*
- The only apes which through *special satellites* (Hubble, COBE, WMAP...) have been able to explore the last confines of the observable universe.
- It is true, however, that man is the only ape that has often forgotten that it is not exactly the same thing to penetrate the secrets of nature and to consider that he is the owner and lord of nature.

After putting together this rough sequence of singular events in cosmic history some may still insist that the sequence is a succession of *chance* occurrences.

Others, more realistic, would recognize something else: *order* and *purpose.*

References

1. M. Denton, *Evolution: A Theory in Crisis* (Adler and Adler, Bethesda, Maryland, 1986).
2. M. Denton, *Nature's Destiny* (The Free Press, New York, 1998).
3. F. Crick, *Life Itself* (Simon & Schuster, New York, 1981).
4. G. Gonzalez and J. W. Richards, *The Privileged Planet* (Regnery, Washington, 2004), Chapter 10.

Chapter 16

A Brief Outline: World Events and Cosmological Discoveries from −4500 to 2010

c.−4500:	Egyptian calendar.
c.−3000:	Astronomical observations in Egypt, Babylonia, India and China.
c.−2500:	Cheops pyramid constructed.
c.−2000:	Equinoxes and solstices determined in China.
c.−2000:	Abraham, Isaac, Jacob: Promise of the Messiah.
c.−1500:	Moses: The Ten Commandments.
c.−1500:	Advanced shipbuilding in Mediterranean and North Sea.
c.−800:	Babylonian and Chinese planetary tables.
c.−550:	Thales of Miletus predicts solar eclipse.
−480:	Buddha dies: His teaching spreads through India and all Asia.
−479:	Confucius dies: His teaching spreads through China and Korea.
−427/−323:	Plato (−427, −347), Aristotle (−384, −322), Alexander the Great (−356, −323)
−323:	Euclid's *Elements*

−250:   Aristarchus of Samos: First heliocentric proposal.
−240:   Erathostenes: Accurate estimate of radius of
        Earth.
−140:   Hipparchus: Improved astronomical observations.
        Discovery of the star "Novae". Trigonometry.
 −46:   Adoption of Julian calendar in Roman Empire.
  −7:   Probable date of Jesus Christ's birth in
        Bethlehem.

  33:    Probable date of Jesus Christ's crucifixion in
        Jerusalem.
  64:    Peter and Paul executed in Rome: Church
        spreads through Roman Empire.
  94:    Trajan: Roman Empire reaches greatest
        geographic extension.
170:    Ptolemy: Geocentric planetary system (deferents
        and epicycles).
600:    Book printing in China.
622:    The Hijrah: Muhammad's flight from Mecca to
        Medina.
        Origin of Muslim calendar. Islam spreads through
        East and West.
828:    *Almagest*: Translation of Ptolemy's planetary
        tables to Arabic.
975:    Arabic arithmetic notation introduced in
        Medieval Europe.
1099:   First Crusade: Christians conquer Jerusalem.
1100–1300:   Medieval European Universities: Bologna (1119),
        Paris (1150), Oxford (1167), Salamanca
        (1217), Cambridge (1200) *etc.*
1200:   Saladin's army recovers Jerusalem.
1211:   Genghis Khan invades China.
1330:   Jean Buridan (1295–1358) teaches
        "impetus" theory in Paris.
        Inertial motion.
1450:   Fall of Constantinople to the Turks.

1492:   Columbus's first voyage to New World.

1498:   Vasco da Gama: Sea route to India surrounding Africa.

1500:   Juan de la Cosa: First World Map.

1522:   Magellan and Elcano: First circumnavigation of the world.

1529–1533:   Division of Christendom: Luther (1483–1546), Zwingli (1484–1531), Calvin (1509–1564) break with Church.

1543:   Copernicus: "De revolutionibus orbium celestium". Heliocentric proposal for planetary system.

1570:   Battle of Lepanto.

1583:   Adoption of Gregorian Calendar under Pope Gregory XIII.

1598:   Tycho Brahe: Accurate measurements of planetary motions.

1610:   Galileo Telescope. "Nuntius Sidereus". Copernican planetary system defended.

1618:   Kepler: "Harmonices Mundi". Laws of planetary motion.

1675:   Römer: Measurement of finite speed of light.

1687:   Newton "Principia". Laws of inertial motion. Gravitation.

1764:   James Watt: Invention of heat engine.

1780:   American revolution.

1789:   French revolution.

1820:   W. Herschel: Catalog of 5000 nebulae in the northern hemisphere.

1829:   Non-Euclidean geometries: Gauss (1777–1855), Lobachevski (1793–1856), Bolyai (1802–1860).

1836:   F. W. Bessel: Parallax of 61 Cygni measured.

1842:   Mayer: Principle of conservation of energy.

1873:   Maxwell: Maxwell's equations.

1900:   Planck: Black body radiation law.

1914–1918:   World War One.

    1917: Russian revolution.
1916: Einstein's General Relativity. Cosmological equations.
1922: Friedmann's solutions to Einstein's equations.
1927: Lemaitre's solutions to Einstein's equations (independently) Big Bang model for a expanding universe.
1939–1945: World War Two.
1948: Alpher and Herman predict cosmic background radiation (CBR).
1965: Penzias and Wilson detect CBR.
1989: COBE: Planck's spectral distribution of CBR measured (Mather). Minute anisotropies in CBR measured (Smoot).
2003: WMAP: Improved measurements. "Age" of universe precisely estimated: 13.7×10^9 yrs. Big Bang basically confirmed.

We may conclude this section and this book with the following eloquent quotation:

The system to which we have likened the universe is an intelligible system. It contains no irrationalities save the one supreme irrationality of creation — an irrationality indeed to physics, but not necessarily to metaphysics... Those who feel the question ["Why the universe?"] to be a permissible one, can legitimately answer the question "why?" by positing God.

The physicist and cosmologist then need God only once, to ensure creation. For the biologist the world provides further opportunity for divine planning, if we admit the possibility of not entirely coincident evolutionary trends in similar circumstances. For man as more than cosmologist, as more than biologist, as possessing mind, possibly endowed with an immortal soul, God is perhaps needed always. Theoretical cosmology is but the starting-point for deeper philosophical inquiries".

E. A. Milne, in *Relativity, Gravitation and World Structure* (Clarendon Press, Oxford, 1935), p. 140.

Appendix

Physics and the Universe: From the Sumerians to the Late-Twentieth Century

Stanley L. Jaki

The monumental stillbirths which science suffered in all great ancient cultures, including the Greeks, is traced to their commonly shared pantheistic view on the universe. Within that view it was impossible to develop a science (physics) which can deal with that chief characteristic of things, which is their being in motion. The breakthrough to a viable science (physics) came in terms of a reflection on the reality of a universe created out of nothing and in time. Although in modern times this Christian notion of the universe has been largely abandoned, the best developments in science, though not in philosophies cultivated by many leading scientists, imply an endorsement of that very view.

Originally, I planned to speak on physics and the universe in the twentieth century. Professor Julio Gonzalo, the organizer of this Conference, had a different idea which, I presume, was suggested to him by my book, *Science and Creation: From Eternal Cycles to an Oscillating Universe*. In the first part of that book, first published in 1974,[a] I analyzed in seven chapters what I call the stillbirth which science suffered in all major ancient cultures, such as the Hindu, Chinese, Egyptian, Babylonian, and the Greek.

No chapter in that book was on the Sumerians. In fact, I hardly mentioned them even in connection with the Babylonians. Nevertheless, Prof. Gonzalo insisted that I begin with the Sumerians even if I talk about physics and the universe. Of course, one can gain a profound insight about the connection between the science of physics and the physical universe by looking at that connection in a historical perspective. But what about the Sumerians? The universe certainly existed at the time of the Sumerians, but physics they did not have. To give the benefit of the doubt to Professor Gonzalo, I went to the Princeton University Library and took out two books on the Sumerians. One of them seemed particularly promising as it contained a chapter on Sumerian scientists.[b] Only science, and certainly the science of physics, was missing in that chapter.

Why then do I still begin with the Sumerians? Not entirely out of courtesy for Prof. Gonzalo. I do so out of the conviction that the absence of science, and of physics in particular, among the Sumerians provides a dramatic background, a sort of indirect proof, on behalf of the thesis I am setting forth in this lecture.

In fact I can no longer say that I could care less for the Sumerians. At one point, around 1970 or so, I certainly thought that I shall always leave them alone. As I said, I mentioned them but fleetingly in my book, *Science and Creation*, a fact that caused a two-year delay, if not more, in the publication of that book. A publishing house in the USA was more than ready to take it. Indeed they were so eager as to ask a Nobel-laureate neurophysiologist, who had two years earlier given them a glowing report about my *Brain, Mind, and Computers*, to evaluate for them the manuscript of *Science and Creation*. But this time the outcome was different. The reviewer wrote a fairly long report in which he argued that I should speak at length about the Sumerians and, moreover, that I should make the Sumerians the subject of the very first chapter in my book. I demurred and had to look for another publisher. By the time the book was typeset there came the energy crisis of 1973. As a result the book was published not in 1971 but in 1974. I had reason for not being enthusiastic about the Sumerians.

Then around 1976 I attended a lecture given by the same neurophysiologist in Philadelphia. His principal thesis was the creativity of the mind, a creativity particularly evident in the arts from the earliest times. Most in the audience, including myself, had not seen before what his first slide showed: a beautiful life-sized alabaster head of a woman. My first reaction was to take it for an early Renaissance work. Amazement ran through the hall as he identified the sculpture as one made in Sumer sometime around 2400 B.C.

Here was a case of an indisputably modern view of the human head: expressive, gentle, almost sweet in spite of its dignified composure. 2000 years later Greek sculptors were just beginning to part with the standards of archaic rigidity and frozen smiles. 4500 years ago the Sumerians could display genuine modernity and in the arts where it seems to count most. The Egyptians had just started building pyramids but for a long time they did not yet sculpt or paint a female head as appealing as that Sumerian sculpture known now as the Warka Head. The Babylonians who largely inherited the Sumerian culture never came up with anything similarly close to the modern spirit.

The Babylonians continued, of course, with the two chief Sumerian inventions: the wheel (used also as potter's wheel) and cuneiform writing. The latter is a great misnomer because it makes us focus on the incidental and not on the essential. The incidental is the wedge-shaped trace left in the clay by each stroke of the iron pen used by the Babylonians. The essential is that all those wedge-shaped traces stand for letters. Now that writing has been around for 4000 years, it is difficult to appreciate what a marvel of abstraction is any letter of the alphabet. In this technological age, with instantaneous transmission and storage of images, there remains little appreciation for a far greater discovery. It is the transformation of phonetic data into abstract visual figures and their combination into groups representing words, or logoi, that are at the basis of all logic in particular and all reasoned discourse in general.

The very word, logic, should take us to the Greeks of old. They were the masters of logic in two respects: One is the development of formal logic in Aristotle's *Prior and Posterior Analytics*. The other is geometry. Euclid's fourteen books do not need our praises. Some of his propositions in the 13th and 14th books are still a challenge to first-rate mathematicians. The Greeks could be startlingly modern in other respects as well. It is enough to think of their dramatists, political theorists, and historiographers. But the Greeks, to say nothing about other great ancient cultures, failed to become modern in one very modern field, which is science.

To a somewhat less startling measure this is a fact also about the ancient Hindus, Chinese, Egyptians, and Babylonians. They all abounded in technical and speculative talent. They all were great organizers, they all interacted with other cultures, they all had material needs. But the impact of none of these factors, which are attributed decisive importance nowadays, availed them with respect to science. The Chinese of old were the inventors of the magnet, of gunpowder, and of block printing — three factors to which Francis Bacon mistakenly attributed the birth of science in early modern Europe.[c] The ancient Hindus devised the decimal system of counting, including the positional value of zero, a stunning discovery. The Egyptians of old built pyramids without pulleys and excelled in practical land surveying. Last but not least, they co-invented with the Sumerians phonetic writing.[d] The Babylonians of old compiled a vast set of astronomical observations that enabled Hipparchus to discover the precession of equinoxes.

Of course, the Greeks were most original in systematizing ideas about lines, figures, and bodies into a set of interconnected propositions. In doing so they invented the science of geometry. In more than one way geometry enabled the Greeks to find out most important things about the physical world. It should be enough to think of Eratosthenes' method of calculating the size of the earth, and of the even more ingenious method whereby Aristarchus of Samos calculated the relative distances of the earth, the moon, and the sun from one another and their absolute sizes and distances as well. Geometry worked marvels in the hands of Ptolemy whose system was as accurate in predicting the position of planets as was the system of Copernicus.[e]

While Archimedes' famed expertise in geometry served him well in working out the basic laws of hydrostatics, he is not known to have done any work in hydrodynamics. For that reason alone one may suspect that Galileo relied on much more than geometry as he specified (on the basis of a long-standing late-medieval doctrine[f]) the spaces covered in equal times by freely falling bodies or the corresponding distances on an inclined plane. Clearly, the Greeks did not possess something as they came to grips, time and again, though in vain, with the reality of motion, embodied either in the free fall of bodies or their traveling through the air as projectiles.

That something relates to the difference between two kinds of science — one that can cope with problems of statics, that is, with things on which no force acts, and the other that deals with problems of dynamics. Therein lies a principal difference between ancient times and modern times and also the clue to the failure of the Greeks of old to become truly modern. Although the same is true of other ancient cultures, in the case of the Greeks the problem can be analyzed with great precision.

Here too Aristotle provides the most instructive case. About his physics it was correctly said that it contains as many errors as pages. His errors are at times astonishing both because they are the very opposite to what Newtonian physics says and because they contradict most obvious everyday evidences. According to Aristotle the time of the fall of a body is inversely proportional to its mass. Aristotle specifically states that if two bodies are released from the same height, the one with twice of the mass of the other will reach the ground in half of the time which it takes for the other body to do so. Aristotle also states in the same context that this is so because the twice heavier body has twice as great desire to rush towards its natural place, the center of the earth.[g]

As in almost all other respects, here too, Aristotle, who loves to recall the importance of first principies, is very consistent. For him the fall of a body is but a particular case of a much vaster and more fundamental motion, or the motion of the starry heavens which in turn has its source in the very nature of the universe as he understands it. Of course, when Aristotle says that the universe is the foremost living entity,[h] he is far from that crass animism that can be found among the Hindus, the Egyptians, and the Chinese of old, and even in the writings of some ancient Greek philosophers, especially some Neoplatonists.[i] Still the universe, as seen by Aristotle, is a subtly animated being. Its animation begins with the desire of the sphere of fixed stars for the Prime Mover, which is really not moving anything. The Prime Mover of Aristotle is not really distinct from the Empyrean Heavens, or the mythical realm above the sphere of the fixed stars.

Some interpreters of Aristotle, Thomas Aquinas, for instance, would disagree on this evaluation of the Aristotelian Prime Mover. However that may be, the desire of the heavenly sphere is, for Aristotle, a sort of divine affection because it is eternally the same. This eternal sameness is manifested in the uniform circular motion of the heavens. But as that desire is transmitted downward it gradually loses its perfection. The name, planets (or vagabonds), is an expression of this cosmic view in which a thing is the more imperfect in its motion the farther it is from the heavenly sphere. Below the moon's orb no motion can be truly regular. Consequently, Aristotle contradicts himself (as well as the facts) as he makes the rate of fall of bodies on earth a function of their mass.

In a sense Aristotle is inconsistent even when he endorses what is obvious, namely, that the fall of bodies on earth is invariably vertical. For him the reason for this is that bodies on earth share, by moving toward the center of the earth, in the perpetual trend of the universe toward self-fulfilment. But this sharing has to be very imperfect, an imperfection that rebounds on the universe itself. This can be seen when Aristotle unfolds the consequences of his pantheistic view of the universe. With his very clear idea of at least some aspects of divine perfection, Aristotle is forced to deny divinity to the world wherever it shows apparenty irregular changes, which it certainly does in the sublunar or terrestrial regions. Hence the Aristotelian emphasis on the essential difference between the heavenly and terrestrial regions of the universe. The consequences of this dichotomy for physics is that there can be an exact physical science only about the heavenly regions of the universe, but not about its terrestrial parts. In other words, one can calculate exactly the motion of the heavenly sphere, and almost exactly the motion of planets, but it makes no sense to calculate the essentially irregular motion of bodies in the terrestrial realm, that is, below the moon's orbit. A dynamics, valid about the entire universe, is an impossibility, in fact a near blasphemy, as far as Aristotle's system is concerned.

It should therefore be no surprise that Aristotle says the correct things in physics only when he takes up the non-dynamical or static character of bodies, such as the direction of their free fall as necessarily perpendicular everywhere to the surface of the earth. The problem of his thoroughly mistaken dynamics should seem therefore all the more tantalizing, a word most appropriate in this connection. With his torments in Hades, Tantalus is a graphic illustration of the inner logic of the conceptual impact of the universe as conceived by the Greeks. Although eternal and therefore without beginning, that universe can have no true fulfilment. At the very moment when its heavenly movements seem to have completed a full circle (when all the planets, as Plato had suggested, are back at the same point and complete a Great Year), everything has to start all over again. The branch full of apples that hangs over Tantalus is moved by the breeze away from him just whenever he almost grabs it.

In short, the Greeks of old failed to formulate physics because their notion of the universe deprived them from making some basic conceptual breakthroughs. The matter should seem all too clear against the background of Newtonian physics which is based on Newton's three laws of motion. The first, or the law of inertial motion, was in substance formulated by John Buridan, more than three hundred years before Newton. He took his clue from the creation of the universe in time.[j] Both words, creation, and in time, are crucially important. The createdness of the universe meant for Buridan that the universe was not a necessary thing either in its existence or in its form. As a non-necessary entity, the physical universe could be for Buridan (and even more so for his successor, Oresme, at the Sorbonne) a mere clockwork in which the motion of parts was not subject to desires and other animistic factors. Such a universe could also be infinite, with no boundaries for the inertial motion of bodies, a conceptual framework very much in the mind of Descartes, whom Newton took for the first formulator of the law of inertial motion.

Both Descartes and Newton owed their notion of the physical universe to that Christian background which greatly helped Buridan and Copernicus in advancing the cause of physics. In such a universe it was possible to talk, as Newton did, in the same breath about the fall of an apple from a tree and of the fall of the moon in its orbit. In such a universe, not subject to ever variable animistic forces, it was possible to credit a form of force mathematically always identical, equal to the product of mass and acceleration, with the motion of all bodies in the universe. In such a universe, and in such a universe alone, an apple could fall on Newton's head and be a clue to the falling of bodies anywhere. Only in such a universe were great scientific geniuses free of the fate of Tantalus as they kept looking for the fruits of their research.

The story of physics since Descartes and Newton up to the present, is a story of the attitude, conscious or tacit, towards the notion, a most Christian notion, of the universe, which gave physics its viable birth in the first place. As long as nothing has been proposed or assumed, either implicitly or explicitly, by physicists against universe as a philosophical construct, the newly born physics could readily unfold its potentials. Newtonian dynamics could be extended beyond the solar system. The inverse square law was successfully applied in electricity. Heat could be treated as the motion of idealized atoms. Electricity and magnetism were united.

Such was a marvelous progress which raised physics to the highest cultural pedestal. But physics was heading toward a potential straightjacket when certain ideas incompatible with the universe were grafted on it. First came a growing scepticism about the doctrine of the createdness of the universe and of its having been created in time. An early case was Descartes himself. If not in words, at least in intention he championed the idea of a universe which was infinite, had an a priori material structure, and whose start was not so much a creative act on the part of God, but a mere flip, as Pascal already diagnosed Descartes's cosmology correctly.[k] With his a priori preference for unbounded Euclidean extension, Descartes laid the groundwork for an ever more worshipful attitude toward what later became known as the idea of an infinite Newtonian universe.[l] Such a universe began to play the role of a substitute for God from the mid-eighteenth century on for two reasons: If it was infinite, it seemed to make unnecessary that other infinite which is God.[m] And since it was infinite in a way (Euclidean three-dimensionality) which is natural for human perception, it was readily assumed to exist naturally, that is, without a Creator.[n] These were developments of great concern for theology, but here we are concerned with physics.

Physicists could have easily seen that an infinite Euclidean universe was contradictory in character. The optical paradox of an infinite Euclidean universe was openly discussed around 1700 and even in the Royal Society with Newton in the chair and with Edmund Halley as the lecturer. Halley, a well-known atheist, tried to take the sting out of the paradox and to shore up thereby the cause of the infinite universe. As it usually happens in such countertheological maneuverings, Halley too had take back with one hand from the universe what he had tried to secure to it with the other. Infinity could hardly be reconciled with the inhomogeneity which Halley, in all appearance, offered as a solution of the paradox.[o]

Just as the optical paradox was largely ignored throughout the 18th century, so was its gravitational counterpart, although already Bentley called, though only qualitatively, Newton's attention to it.[p] No physicist paid attention when in 1828 Green proposed the theory of potential, on the basis of which a strict formulation can be given to the gravitational paradox. Again, there were no notable reactions when in 1848 the younger Herschel showed that absorption of light in interstellar ether cannot solve the optical paradox. Nobody saw new opportunities for physics when in 1872 Zollner suggested in detail a four-dimensional geometry for the solution of the gravitational paradox. No physicist protested when in 1884 Lord Kelvin defended Euclidean infinity as the only form of rationality.[q] No physicist shook his head in disbelief when in 1901 Lord Kelvin proposed a solution to the optical and gravitational paradoxes on the ground that the infinite number of bright stars beyond the Milky Way was of no consequence for the visible part of the universe which he identified with the Milky Way![r]

Ultimately it was Kelvin's belief in that infinity that made him belittle as two small clouds in the bright sky of physics two problems: one posed by the black body radiation, the other by the null result of Michelson's ether-drift experiment. This worshipful respect toward the infinite Newtonian universe resulted in a schizophrenic view, subtly reminiscent of the Aristotelian cosmic dichotomy. The Newtonian infinite universe was split into two parts: one visible and puny, restricted to the Milky Way, the other the infinite beyond it and forever beyond the investigation of the physicist.[8]

For physics this meant a delay in the advent of General Relativity — for if the optical and gravitational paradoxes of an infinite universe had been kept in focus, a psychological pressure might have built up already in the eighteenth century for a consideration of the possibility of non-Euclidean geometries. It is well known that Gauss did not dare, in the early 1800's, to publish his ideas about non-Euclidean geometries lest he be taken for a fool not only by the world at large but also by the world of mathematicians and physicists. Clearly, the unjustified respect for a notion of the universe, which was wrong scientifically in the same measure in which it was anti-Christian, could but hamper the progress of physics.

No different is the twentieth-century part of the story. On the one hand we have the marvelous physics of General Relativity which assures the greatest discovery in the history of physics. The discovery is the first contradiction-free discourse about the totality of consistently interacting things or the universe. It is on that discovery that rests the rapidly expanding field of modern scientific cosmology in which one sees united the study of galaxies and stars with the study of fundamental particles. Einstein himself was rightly proud of his achievement. He saw the proof of his genius not in Special Relativity which first contained the famous formula mc^2 but in General Relativity with its consequences for cosmology.

Unfortunately for Einstein he viewed the universe as divine in a broad pantheistic sense. He therefore wanted to keep the universe essentially immobile, that is, unchanging. This is why he tried to offset its possible expansion by introducing the cosmological constant. Years later he had to admit that in doing so he committed the greatest blunder of his scientific career. He was not at all pleased when the idea and fact of the recession of galaxies had to be recognized. In order to save eternity for the universe, he then endorsed, in 1931, the idea of an oscillating universe which he had considered with great reservation nine years earlier.[t] Blinded as he was by his pantheistic respect for the universe, Einstein failed to see the obvious, namely, that the oscillations of the universe had to die out eventually.

What happened in terms of quantum theory, tells a similar story of oversight and blindness with respect to the universe. Here the first thing to note is that by 1927, when Heisenberg derived his uncertainty principle, he and a number of other physicists had explicitly rejected causality. They did so on philosophical grounds which for the most part they failed to study carefully. As is well known, around 1900 most physicists subscribed to some mixture of Humean empiricism and Kantian idealism yielding some pragmatist and operationist philosophy in which there was no room for causality in the ontological sense. In fact, some physicists were such poor philosophers at that time as to take statistical gas theory for an overthrow of causality. There is therefore much more than meets the eye in Heisenberg's statement that because of the uncertainty principle, "the principle of causality has been definitively disproved."[u]

What were the consequences for physics? I do not, of course, mean the further unfolding of the marvelous potentialities of quantum mechanics as a scientific theory and technique. In a sense, though not in exactly the same sense as Schrödinger wanted it, one can talk of quantum mechanics as "the Lord's quantum mechanics."[v] Its successes, vast and numerous, are too well known to be recalled in detail. But little has been said about the consequences for physics, for quantum mechanics in particular, of a view of the universe in which the universe is deprived of causality and becomes thereby a mere heap of random occurrences. The chief consequence for physics is that it is turned into a most dubious means whereby one can pull a rabbit from under one's hat, that is, a rabbit which was not there in the first place.

In fact not only rabbits can be "created" this way but the universe itself, and if this were not nonsense enough, several universes, indeed an uncounted number of them. The start of that hideous disrespect for reality is Heisenberg's uncertainty principle insofar as it is taken for a principle of anti-ontology and not for what it truly is, a limit put on the precision of measurements. When Heisenberg stated that the principle of uncertainty definitively disproved causality, he spoke not as a first-rate physicists but as a third-rate philosopher. He simply overlooked that what he proposed was kind of elementary fallacy for which the Greeks of old already had a technical name: *metabasis eis allo genos*. For what Heisenberg claimed, and countless physicists after him, is a jump, a metabasis, from the purely operational to the strictly ontological domain. As such the jump or move is a plain non-sequitur.

It can be rephrased as follows: An interaction that cannot be measured exactly with the tools, conceptual and instrumental, at our disposal, then such an interaction cannot take place exactly. The non-sequitur follows from taking the same word exactly, first in the operational and then in the ontological sense. Such a procedure is bordering on plain equivocation.[w]

If, however, one accepts that philosophical fallacy, one will have, in terms of Heisenberg's uncertainty principle, a pseudo-scientific proof of the foremost breakdown of causality. Such a breakdown occurs when one speaks of the creation of matter out of nothing but without a Creator. The first to graft this skullduggery on physics were the steady state theorists. They did so in support of their explicit belief in the universe as the ultimate entity, a sort of a substitute god. Like Einstein, they too wanted to save the universe from an overall change, from a possible ultimate dissolution into a practically zero density of matter. Exceedingly small was the amount they claimed to pull out of nothing into existence in order to keep the universe the same or unchangeable as befits a god. One hydrogen atom per second in every cubic mile of space is exceedingly small. But since there are an astronomically large number of cubic miles in the universe, the total amount of hydrogen thus "created" amounted to astronomically large quantities.

Revealingly, the scientific community did not attack the steady-state theorists for their misuses of philosophy. Countless physicists let themselves be pulled into the belief that the basic postulate of steady state theory has scientifically verifiable consequences, such as the extra amount of 21 cm radiation, the natural wavelength of the hydrogen atom. But even if that extra amount had been detected, it would not have proved the steady state theory. That basic postulate of the theory is beyond scientific proof or disproof. There can be no scientific proof of creation out of nothing, be that creation by God or by those parading as scientific demigods. The elementary proof of this is that the nothing is by definition unobservable even with the most refined scientific instrument.

Time does not allow me to trace here, however briefly, the development from the steady state theory to the inflationary theory and to the multiworld theory of the universe. In both, ability is claimed by physics to create universes literally out of nothing. Nor is there time here to review the anti-theistic inspiration of the inflationary theory of the universe in which the universe, this very object that gives meaning to physics, is being fragmented into an immensely large number of "universes," all disconnected from one another by the statistically different sets of laws governing each.

Here only a few words ought to be said about the impact on physics of such unconscionable games with the universe. The impact is a decrease of awareness of the possible shortcomings of quantum theory. For if quantum mechanics is taken by the physicist for an assurance that, to quote a professor at MIT, he can create universes literally out of nothing,[x] then he will hardly pay attention to the warning which Dirac gave at the Jerusalem Einstein conference in 1979[y]:

> It seems clear that present quantum mechanics is not in its final form ... Some day a new relativistic quantum mechanics will have determinism in the way that Einstein wanted. This determinism will be introduced only at the expense of abandoning some other preconceptions which physicists now hold, and which it is not sensible to try to get at now ... So under these conditions I think it is very likely, or at any rate quite possible, that in the long run Einstein will turn out to be correct even though for the time being physicists have to accept the Bohr probability interpretation — especially if they have examinations in front of them.[y]

I did not quote Dirac in order to suggest that it is the fault of quantum mechanics that it cannot cope with General Relativity. It may be the other way around. Quantum mechanics, as it stands, is a wonderful physics, though limited by its very basic assumption, namely, that atomic processes can only be dealt with in groups, and not invididually, and only in terms of a statistical theory steeped in non-commutative algebra. Indeed, as Aspects's experiments showed, Bell's theorem is correct in predicting that on the atomic level quantum mechanical statistics fits the phenomena better than does classical statistics. This is, however, merely a proof that quantum mechanics, as it stands, cannot be complemented with a theory of hidden variables, but not a disproof of the theoretical possibility of a new form of quantum mechanics as suggested by Dirac.

But neither quantum mechanics, nor any form of physics, however successful, can assure the physicist that he possesses the last word about the physical universe. As an experimentalist, the physicist can never be sure that the universe has no more hidden aspects to it. As a theoretician, the physicist should be mindful of Godel's incompleteness theorems which forclose the derivation of the consistency of an all-inclusive physical theory on a priori grounds.[z] Yet the physicist must assume the existence of a universe, fully consistent in its operations and fully investigable scientifically, if the science of physics has to retain its meaning.

About that universe no physics, Newtonian or modern, provided deeper insight than the one offered by theology. If we turn to John Henry Newman in this centenary of his death for such an insight, we will not be disappointed. In his *The Idea of a University* he wrote: "There is but one thought greater than that of the universe, and that is the thought of its Maker."[aa] And if we turn to that theologian, Thomas Aquinas, whose timeliness depends even less on anniversaries, we can hear the same praise of the universe in terms even more startling: "God himself could not have created something greater than the universe."[ab] Only such an exalted view of the universe can keep physics in a safely exalted position.

Endnotes

a A second enlarged edition was published, also by Scottish Academic Press, Edinburgh, in 1986 and reissued in the USA by the University Press of America, Lanham, Md., in 1990.

b i. E. Lansing, *The Sumerians: Inventors and Builders* (McGraw-Hill, New York, 1978).

   ii. L. Cottrell, *The Quest for Sumer* (G. P. Putnam's Sons, New York, 1965), the handling of Sumerian "science" consists in references to the "science" of Babylonians of Hammurabi's time. Both works heavily depend on.

   iii. S. N. Kramer *History Begins at Sumer* (1959).

c S. L. Jaki, *The Origin of Science and the Science of Its Origin* (Scottish Academic Press, Edinburgh, 1978), pp. 7–16, on the thoroughly mistaken ideas of Bacon on the origin of science.

d This point is made because of the widespread custom of taking Egyptian hieroglyphics for a form of ideographic writing, similar to the one still used in China.

e The reference is, of course, to the system of Copernicus himself and not to the subsequently much improved forms of heliocentric planetary theories.

f The doctrine in question is known as the impetus theory, first proposed by John Buridan at the Sorbonne around 1330 and discovered by Pierre Duhem around 1906 through his pioneering study of medieval manuscripts. By the time Copernicus relied on the impetus theory to solve the physical problems arising from the twofold motion of the earth, the impetus theory had been so generally accepted as to make unnecessary references to its original formulator.

g See *On the Heavens*, pp. 273b–274b.

h See *On the Heavens*, p. 285a.

i A particularly glaring example is *De antro nympharum* of Porphyry (233–C.304) disciple and editor of Plotinus.

j S. L. Jaki, *The Savior of Science* (Regnery Gateway, Washington D. C., 1988), pp. 52–53.

k "I cannot forgive Descartes. In all his philosophy he would have been quite willing to dispense with God. But he had to make Him give a fillip to set the world in motion; beyond this, he has no further need of God." B. Pascal, *Pascal's Pensees,* translated by W. F. Trotter, with an Introduction by T. S. Eliot (E. P. Dutton, New York, 1958), p. 23.

l The idea was un-Newtonian in that Newton held the physical universe to be finite, surrounded by a no less finite layer of the ether, beyond which was the infinite, though non-physical, space, taken by Newton for the "sensorium of God."

m Such a reasoning, cheerfully accepted by agnostics in the nineteenth century, still caused perplexities to eighteenth-century deists, as is clear from the article "Infini" in the *Encyclopedia* of Diderot and d'Alembert, where it is clearly stated that only God can be strictly infinite.

n The same reasoning reappeared in our time in the cosmology of steady-state theorists and, more memorably, in the concluding chapter of S. Hawking, *A Brief History of Time.* See S. L. Jaki, Review of Hawking's book, "Evicting the Creator," in *Reflections* (Spring 1988), reprinted in S. L. Jaki, *The Only Chaos and Other Essays* (University Press of America, Lanham, Md.; Intercollegiate Studies Institute, Bryn Mawr, PA., 1990), pp. 152–161.

o For details, see S. L. Jaki, *The Paradox of Olbers' Paradox* (Herder and Herder, New York, 1969), Chapter 4.

p S. L. Jaki, *The Paradox of Olbers' Paradox,* pp. 60–62.

q To make matters worse, Lord Kelvin did so in the context of a much publicized lecture in Philadelphia. See S. L. Jaki, *The Paradox of Olbers' Paradox,* p. 170.

r For details, see S. L. Jaki, *The Milky Way: An Elusive Road for Science* (Science History Publications, New York, 1972), pp. 276–277.

s The topic of Chapter 8, "The Myth of One Island," in S. L. Jaki, *The Milky Way.*

t For details, see S. L. Jaki, *Science and Creation,* p. 357.

u The phrase is from Heisenberg's paper of 1927 containing the formula for the principle of uncertainty.

v E. Schrodinger, *What Is Life and Other Scientific Essays* (Doubleday, Garden City, New York, 1956), p. 83.

w A point on which I have repeatedly insisted for now over a dozen years, most recently in S. L. Jaki, Determinism and reality, *Great Ideas Today 1990* (Encyclopedia Britannica, Chicago, 1990), pp. 278–302.

x The professor in question is A. H. Guth, who made this preposterous claim both in published interviews and in scientific publications. For details, see S. L. Jaki, *God and the Cosmologists* (Regnery Gateway, Washington, D.C., 1989), p. 258.

y Dirac statement was quoted by R. Resnick, a participant in that Conference, in his article, Misconceptions about Einstein: His work and his views," *Journal of Chemical Education* **52** (1980) 860.

z For a detailed discussion, see Chapter 4 "Godel's shadow" in S. L. Jaki, *God and the Cosmologists.*

aa See the 8th edition of S. L. Jaki, *My God and the Cosmologists* (Longmans, Green and Company, London), p. 462.

ab For details, see S. L. Jaki, Thomas and the Universe, in *The Thomist* **53** (1989) 545–572.

Glossary

Alpha particle. Nucleus of ^{4}He atom made up of two neutrons and two protons. After primeval nucleosynthesis the cosmic plasma is made up mainly by protons (76%), α particles (24%) and electrons.

Baryon-to-photon ratio. At present the universe is *transparent* and the ratio of baryons to photons in the universe is $n_b/n_r \approx 10^{-9}$ and is kept constant. It is to be expected that before atom formation, at temperatures above 3000 to 4000 K, the universe was in an *opaque* plasma state, in which there was considerable friction (multiple scattering) and the density of photons increased with time with respect to the density of baryons (nuclei of H and ^{4}He) (See Chapter 12).

Baryogenesis. Process by which a definite population of baryons (protons, neutrons) becomes well defined in the early universe. This requires that the cosmic matter density becomes equal or less than nuclear density, which occurs at a cosmic temperature $T_b \approx 3.88 \times 10^{12}$ K. It must have taken place at $t \sim 10^{-5}$ s.

Big Bang nucleosynthesis. Primordial process by which cosmic protons and neutrons fuse together to form ^{4}He nuclei and traces of other light nuclei. This occurs at a temperature $T_{ns} \sim 4.60 \times 10^8$ K. It must have taken place at $t \sim 10^3$ s.

Blueshift. If a star or galaxy moves toward us the radiation emitted by it observed from the Earth appears shifted toward shorter wavelength (towards the blue) (See Doppler shift).

Cepheid variable. A type of variable star whose brightness was seen to vary over time. Henrietta Swann Leavitt discovered in 1912 that an approximately linear relationship existed between the period of a Cepheid's pulsation and its luminosity. The longer the period of its pulsation, the greater was the star's luminosity. Once the distance to nearby Cepheids was determined, they became useful measuring sticks for establishing the distances to other galaxies after they were discovered. In 1925 Edwin Hubble discovered a Cepheid in M31, the Andromeda Nebula, which enabled him to estimate that it was 8×10^5 light-years from Earth. It convinced astronomers that Andromeda was not part of the Milky Way, but a galaxy in its own right.

Cosmic accelerated expansion. Recent observations of magnitude (related to distance) versus redshift (related to recession velocity) which indicate that the rate of change of velocity with distance is larger for nearby galaxies than for far-away galaxies. To properly interpret the evolution in rate of change it is necessary to take into account that the maximum recession velocity cannot exceed the speed of light and that for the most distant protogalaxies relativistic effects must be expected (See Chapter 10).

Cosmic time parameter $H \times t$. Cosmic parameter (dimensionless) giving the product of the Hubble's parameter $H = \dot{R}/R$ and the cosmic time at any moment in the cosmic expansion. For an open universe $H \times t$ is less than one and more than two third (2/3). This parameter is time dependent, because $H \times t$ is time dependent and, of course, t is also time dependent. At present, using local data, $H \cdot t \approx 0.94 \pm 0.06$, substantially larger than $H_{af}t_{af} \approx 0.667 \approx 2/3$, corresponding to the time of atom formation ($t_{af} \approx 300.000$ yrs).

Dark energy. Since $\Omega \approx 1$ is consistent with the estimated matter-energy density at the time of "decoupling" (T $\approx 3\,000$ K), and the dark mass has been taken as contributing about 30% to it, the difference (about 70%) is commonly attributed to the repulsive force of the vacuum (cosmological constant) in the form of a kind of potential "dark energy": For a discussion of the time dependence of the matter mass density ($\Omega_m(t)$) and the potential space-time curvature energy density ($\Omega_k(t)$) resulting in $1 = \Omega_m(t) + \Omega_k(t)$, compatible with

$\langle \Omega_m \rangle \approx 0.26$, $\langle \Omega_k \rangle \approx 0.74$, averaged from the time of galaxy formation to the present, see Chapter 11.

Dark matter. In the close neighborhood of our galaxy the matter mass density in galaxies is only about 4% of the critical density (required for galaxies being exactly at *escape* velocities). It is commonly assumed that in order to have $\Omega = \rho/\rho_c = 1$. Over cosmic space-time a matter mass density of at least 30% is required, including matter of non-baryonic nature. This is the so called "dark matter". In discussing dark matter, the time dependence of $\Omega(t)$ from very early times to present as given by the Friedmann–Lemaitre solutions of Einstein's cosmological equations, is usually ignored (see Chapter 11).

Dark night paradox. The paradox which points out the apparent impossibility of having a dark night in a universe with an infinite number of luminous stars homogeneously distributed.

Doppler shift. Shift in the receiver frequency (and wavelength) of waves (sound or electromagnetic radiation) due to the relative motion of source and observer, depending on their approach or recession. For light, i.e., electromagnetic radiation, results in a blueshift or a redshift, respectively. Christian Doppler first proposed this effect in terms of the change in pitch for sound waves in 1841, but modern astronomers apply it to electromagnetic waves, corresponding to an observed change in color. Stars or galaxies with spectra shifted to the blue (or shorter) wavelengths of the spectrum are moving toward the Earth. Stars and nebulae with spectra shifted to the red, or longer, wavelengths are moving away.

Electromagnetic interactions. Interactions due to electric charges. Static charges give rise to electric fields, accelerated/decelerated charges produce electromagnetic waves, including radio, microwave, infrared, visible, ultraviolet light and X-rays, as well as γ-rays.

Electroweak interactions. Unified description of weak interactions (responsible for nuclear β decay) and electromagnetic interactions, due to Glashow, Weinberg and Abdus Salam (1967–1970).

Energy conservation. Formulated by nineteenth century physicists (originally by Julius R. Mayer in 1842). It is the principle that establishes the strict *conservation* of energy: energy is not destroyed or created in physical processes, it is only transformed, for instance, thermal energy into mechanical energy, or vice versa. After the advent of Relativity Theory, associating an energy $E = mc^2$ to any mass m, the principle was generalized to include the energy associated to rest mass.

Expansion of the universe. The expansion of cosmic space suggested by Einstein's field equations of general relativity. Alexander Friedmann showed in 1922 that such an expansion was a natural consequence of the field equations, but Einstein, at first disagreed with Friedmann, who died in 1925.

Finiteness of the universe. We assume the universe to be intelligible (otherwise we would not be investigating it). It is *infinite* or *finite*. If it is infinite it is beyond our capability of understanding. So we *conclude* it is finite, and hope not to incur a contradiction, as pointed out explicitly by Georges Lemaitre, and as required explicitly by Einstein.

Hubble parameter. The ratio of recession velocity of galaxies ($\dot{R}$) to distance (R), improperly called Hubble's constant, because for nearby galaxies $H_o \approx \dot{R}_o/R_o$ is approximately constant, but, in principle, it is time-dependent, being $H(t) \approx \dot{R}(t)/R(t) \gg H(t_o)$ for early times. The presently accepted value of the Hubble parameter is $H_o \sim 67(\text{km/s})/\text{Mpc}$, with $1\,\text{Mpc}$ (Megaparsec) $= 3.26 \times 10^6$ light years.

Hubble time. Estimate for the age of the universe based on the reciprocal of Hubble's parameter. At the time Lemaitre originally proposed his primeval atom theory, the Hubble time was estimated to be only about two billion years. Later revisions of Hubble's distance estimates increased the value upward to 4×10^9 years in 1948 and upward to 1×10^{10} by the time Allan Sandage took over Hubble's position at Mount Palomar.

Isotropy. Attribute assumed by Lemaitre, and other cosmologists, that the universe appears the same in all directions. To a very high degree, measurements of the cosmic background radiation roughly confirms this.

Leptons. A class of non-strongly interacting particles which includes the electron, the muon, the tau particle, and their associates.

Magnetic monopole. An isolated pole (north or south) of a hypothetic magnet. An infinitely long magnet in which the two poles are so far apart that they do not affect each other would act as a pair of monopoles. Dirac predicted the existence of particles with properties of magnetic monopoles, and so did some grand unified theories. But they have never been observed.

Photon. A quantum (discrete) minimum of electromagnetic energy consisting essentially in a localized particle of light after the quantum theory of radiation put forward by Max Planck in 1900.

Quarks. Elementary constituent of protons, neutrons and other strongly interacting particles (baryons and mesons).

Red giant. A phase in the life cycle of a typical star (not very heavy, like our Sun) in which the size of the star increases tremendously and then blows up into space, leaving as a remnant a white dwarf.

Steady state theory. A cosmological theory postulated by Fred Hoyle, Thomas Gold, and Herman Bondi in Britain in 1948, asserting that the universe is in essence without change. Hoyle, Bondi, and Gold argued against a temporal, Big Bang origin, suggesting that while the universe was expanding, the expansion occurred in a natural state similar to that suggested by de Sitter's universe, but with matter being constantly created in the form of hydrogen atoms in empty space, keeping cosmic conditions identical as a function of time. In the early 1960s the discovery of quasars and other developments in astronomy strengthened the contention that the universe in the distant past was far different from the universe at present. The discovery of the cosmic microwave background radiation all but discredited the state the theory.

Type 1a supernovae. Violent stellar explosions signaling the utter demise of a star, that take place when a white dwarf bleeds enough mass from a companion star to pass a certain mass limit, approximately 1.4 times the mass of the Sun (also called Chandrasekhar's limit, after the Indian-American astrophysicist Subrahmanyan Chandrasekhar), at which point it explodes. The luminance of such supernovae is so great that they can be brighter than entire galaxies as they are photographed. The consistent level of luminosity of Type 1a supernovae have made them an even more reliable distance indicator than Cepheids for determining extragalactic distances.

White dwarf. Final phase in the life of a typical star after it went through the phase of red giant.

Author Index

Alpher, R. A., vii, viii, 53, 72, 73

Bentley, R., 42
Bohr, N., 10, 89, 90
Bolyai, J., 38, 39
Bondi, H., 11
Brawer, R., 53

Cantarell, I., 12
Chandrasekhar, S., 126
Copernicus, N., 9
Cowan, C., 10

de Broglie, L., 90
Dicke, R. H., 19, 27
Dirac, P., 19

Eddington, A. S., 53, 60
Einstein, 39
Einstein, A., 9, 35, 38, 41, 53, 59–61, 87, 88, 90
Engels, F., 87

Friedmann, A., 38, 39, 42, 59, 60, 63, 124

Gamow, G., vii, 60
Gauss, K. F., 38, 39, 87
Gold, T., 11
Gonzalo, viii
Guth, A., 11, 12

Halley, 27
Hawking, S., 90
Heisenberg, W., 90
Herman, R., vii, 53, 72, 73
Hoyle, F., vii, 11, 46
Hubble, E., 42, 47, 53
Humason, M. L., 42

Jaki, S. L., xv, 27, 42, 88, 90

Kepler, J., 9
Kragh, K., 60

Lederman, L. M., 11
Lemaitre, G., vii, 42, 46, 47, 53, 59–63
Lifante, G., xvi
Lightman, A., 53
Lobachevski, N. J., 38, 39

Mach, E., 9, 87
Marx, K., 87
Mather, J. C., viii
Milgrom, M., 72

Newton, I., 9, 10, 38, 42, 88
Nuñez de las Cuevas, R., 12

Olbers, W., 27

Pauli, W., 10
Penzias, A. A., 47, 53, 54
Planck, M., viii, 9, 17, 54, 87, 88

Reines, F., 10
Riemann, G. F. B., 38, 87

Schlipp, P., 61
Schrödinger, R., 90
Schwartz, M., 11
Silk, J., 72
Smoot, G., 12
Solovine, M., 10, 88

Steinberger, J., 11

Vilenkin, A., 12

Whittacker, E., 83
Wilson, R. W., 47, 53, 54

Zwicky, F., 71

Subject Index

^{4}He, 7, 17, 21, 24, 31, 55, 78, 79, 93

13.7 Gigayears (age of universe), 5, 85, 93, 102

a virtual fluctuation of energy, 90
accelerating universe, 65
actual cosmic density, 71
advanced technologies, 97
African eclipse expedition, 60
age of universe, 86
algebra, 97
Altamira, 96
American National Academy of Sciences, 60
American revolution, 101
apes, 97, 98
apparent magnitude, 67
arithmetic notation, 100
astronomical observations, 99
astronomy, 125
atmosphere, 16, 94, 95
atom formation, 7, 22, 50, 55, 75, 78, 122
atomic universe, 7
attractive forces, 15
average values, 75

baryon formation, 55
baryons, 19, 31, 56, 78, 86, 121, 125
Battle of Lepanto, 101
bending of light rays, 39, 40, 60

Big Bang, vii, 4, 5, 11–13, 17, 31, 43, 46, 47, 60, 65, 66, 85, 86, 93, 102
biosphere, 94
Bohr's radius, 21
book printing, 100
bosons, 23

Catholic tradition, 4
CBR, 12, 24, 29–31, 39, 50, 51, 53, 54, 61, 65, 102
    cosmic mass, 19
    dust, 24, 57, 71, 93
    mass, 42
    radius, 18, 25, 38, 42, 45, 46, 57, 66, 69, 71
    time, 79
Cepheids, 126
ceramics, 97
chance occurrences, 98
Cheops pyramid, 99
Christianity, 4
Church, 4, 100, 101
closed universe, 48, 49
COBE, 12, 31, 98, 102
communications, 97
compact parametric solutions, 48
comprehensibility (of the universe), 10
Compton radius, 86
Compton wavelength
    electron, 21
Constantinople, 100

contingent universe, 87
cosmic explosion, 31
cosmic history, 93
cosmic quantities, 18
cosmic times
    electron formation, 80
    end of nucleosynthesis, 80
    neutron formation, 80
cosmological, 71
cosmological constant, 5–7, 12, 37, 38, 45, 60, 62, 65, 73
cosmological implications, 23, 61
counter-gravitational, 42
creation, 3, 11, 102
Creator, 4, 13, 90
Credo, 4
critical density, 71
critical mass density, 48
curvature of space, 29
curved spaces, 42

dark energy, 7, 71, 73, 75
dark matter, 71, 72
    cold, 72
    hot, 72
    warm, 72
dark night sky, 27
decoupling, 7, 22
decoupling time, 65
density parameter
    matter mass Ω_m, 7
    radiation mass Ω_r, 7
density parameters
    cosmological constant Ω_Λ, 6
    space-time curvature Ω_k, 6
dimensionless product, 48
disintegration, 79
distance from the Sun, 95
division of Christendom, 101
domestication of animals, 96
Don Quixote, 97
dust, 7

Earth, 16, 27, 30, 31, 47, 86, 94–97
Earth's location, 96
Earth–Moon system, 94
Egyptian calendar, 99

Einstein
    cosmological constant, 45
    cosmological equation, 25, 39, 69, 91
    general theory of relativity, 28
    letter to Lemaitre, 61
Einstein–Friedmann solutions, 60
electromagnetic interaction, 18
electromagnetic waves, 123
electromagnetism, 97
electron, 7, 10, 17, 21–24, 31, 55, 78–80, 86, 90, 93
elementary particles, 11, 15, 86, 89
empty space, 125
energy
    conservation, vii, 3, 5, 7, 9–11
    density of vacuum, 12
epistemology, 88
equation of state, 7, 49, 50, 55, 73, 78, 80
eternal inflation, 12
evolutionary trends, 102
extraordinarily close to unity, 72

Fiat Lux, 46
fifth interaction, 15
fine-structure constant, 19
finite universe, 88
fireworks event, 46
first approximation, 40
first crusade, 100
first World Map, 101
flat, 56
flat universe, 57
fly to the Moon, 98
four fundamental interactions, 15
free neutron half-life, 79
freedom: too much, 9
French revolution, 101
Friedmann–Lemaitre Solutions, 7, 45

galaxies, 5, 7, 11, 15–18, 24, 27, 29–31, 39, 42, 51, 60, 65–69, 71, 78, 86, 93
galaxy distance and recession velocity, 69
galaxy velocity, 68

GBR, 72
General Theory of Relativity, 37–39,
    41
genome, 95
geocentric planetary system, 100
geodesic, 42
geometric, 37
geometry, 39, 42, 97
Gigayears (age of universe), 51
God, xv, 4, 27, 87, 102
GPS, 98
grammar, 97
Grand Unification Theories, 11
gravitational
    constant, 19, 38, 45
    interaction, 15, 17, 38
gravitational interaction, 38
gravitational mass, 38
great glaciations, 96
Gregorian Calendar, 101

Hamlet, 97
heavy
    metallic elements, 94
    non-metallic elements, 94
Heisenberg
    time, 25, 30
Heisenberg's principle, 90
high energy accelerators, 23
Hijrah, 100
historical records, 97
hubble telescope, 29
hydrogen atom, 10

idealism, 88
Iliad, 97
imaginary worlds, 13
impetus theory, 100
implosion, 31
indeterminacy, 90
inertial mass, 38
inertial motion, 100
inflationary
    factor, 25
interactions, 15
    Strong nuclear, 18
    Weak nuclear, 18

jets, 97
Judeo-Christian tradition, 4
Julian calendar, 100
Jupiter, 94, 96

kinetic energy, 5, 50

lasers, 98
leptons, 23
life's physical preconditions
    amino acids, 96
    chemical properties of C, 96
    DNA, 96
    fortunate locations, 96
    photosynthesis, 96
    protective action of Jupiter, 96
    protiens, 96
    rain of meteorites, 96
    Van Allen belt, 96
    water, exceptional properties, 96
light elements, 24, 94
liquid, 11
logic, 97
logical simplicity, 61

MACHOS, 72
Mars, 94, 95
mass, 85
mathematics, 32, 97
matter, 5–7, 11, 15, 21, 22, 29–32,
    37–39, 42, 46, 48–51, 55, 57, 61, 65,
    71–75, 78, 80, 85, 93, 121, 122, 125
matter pressure, 56
mechanics, 97
medieval European universities, 100
Mercury, 93
metric tensor, 37
Milky Way, 29, 31, 60, 93, 122
miracle (comprehensibility of the
    world), 88
missing mass, 71
Monist religion, 3
motion, 123
Mount Wilson Observatory, 42
Mount Wilson, California, 60
multiverses, 32
muon-neutrino, 11

NASA, 5, 12, 95
natural laws, 12
natural unit, 18
negative pressure, 60
negatively curved distortions, 39
neo-Kantians, 87
Neptune, 94
neutrino, 10, 16, 23, 72
neutron
    half-life, 17
    mass, 17
neutron-to-proton
    calculated, 79
neutron-to-proton ratio
    observed, 80
neutrons, 10, 16, 23, 24, 31, 72, 79,
    93, 121, 125
Newton's constant constant, 61
nineteenth century physics, 3
nitrogen, 94
non-baryonic matter, 72
non-Euclidean geometries, 87, 101
nonlinear partial differential
    equations, 37
nucleons, 16, 23
nucleosynthesis, 31, 79, 93, 121
number of baryons, 55
    galaxies (in universe), 29
    photons, 55
    stars (in galaxies), 29

objective world, 10
Oklo river, 19
Olber's paradox, 27
open universe, 42, 56, 72
optics, 97
order and purpose, 98
origin of the universe, 43
our galaxy (location in universe), 66
oxygen, 94

paradox, viii, 27, 28, 31, 32, 65
Parallax of 61 Cygni measured, 101
particle physics, 12
Periodic Table, 94
phase transition, 11

philosophical inquiries, 102
photon-to-baryon ratio paradox, 77
photons, 77, 78, 90, 93
physical laws, 38
physical meaning of k, 57
physics, ix, 4, 11, 22, 28, 31, 59, 86,
    89, 90, 97, 102
Piccard's lab, 60
Pius XII, 46
Planck
    blackbody radiation, 54
    monopole mass, 18
    time, 91
Planck's units, 18
planetary tables, 99, 100
plasma phase, 55
plasma universe, 7, 78
Pontificia Academia of Sciences, 46
positively curved distortions, 39
positivism, 88
potential energy, 5, 50, 73, 74, 80
primeval atom, 43, 61
primitive men, 96
primordial matter, 24
primordial nucleosynthesis, 24, 55
principle of equivalence, 38
Privileged Planet, 96
process, 121
Promise of the Messiah, 99
protogalaxies, viii, 29, 31, 73, 75
proton mass, 17–19
protons, 16, 21, 23, 24, 31, 55, 72, 79,
    121, 125

Quantum Mechanics, 13, 89
quarks, 23, 24
quasars, viii, 29, 66

radiation, 5, 7, 21, 22, 24, 29–31, 39,
    40, 42, 47, 48, 50, 51, 53–56, 61, 65,
    74, 78, 80, 93, 96, 101
    cloud, 5, 24
    electromagnetic, 21
radiation pressure, 57
radiation-to-matter pressure, 56
radioactive nuclei, 94

radius, 12
realistic metaphysics, 88
recession velocity, 29, 66–68
red giant, 125
redshift, 39, 51, 66, 67
repulsive forces, 15
Revelation, 4
rhetoric, 97
Ricci tensor, 37
Riemann curvature tensor, 37
RNA, 95
Russian revolution, 102

Saturn, 94
Schwarzschild radius, 46, 49, 86
scientific learning, 97
shift in cultural reference, 90
singular events in cosmic history, 98
singularity, 25, 91
Sistine Chapel, 97
solar system, 31, 93–95
solid, 15, 21
Solvay Conference (1927), 59
space-time curvature, 5, 38, 39, 45,
    50, 51, 61, 62, 73–75, 80
speed of light, viii, 29, 31, 38, 40, 45,
    61, 66, 67, 97, 101
spherical homogeneous and isotropic
    universe, 38
spin, 23
stable nuclei, 94
stable atoms, 94
standard model, 23
standing cosmic paradoxes, 73
stars, 7, 16, 17, 24, 27, 29–31, 42, 51,
    72, 78, 93
static universe, 5, 38
Steady State Theory, vii, 11, 12, 31
Stefan–Boltzmann law, 55
subatomic particles, 22, 23
Sun, 5, 15, 16, 21, 29–31, 39–41, 60,
    96, 125
supercooling, 11
superposition, viii

tau-neutrino, 11
Ten Commandments, 99
theory of
    multiverses, 13
    quanta, 9
    relativity, 9, 60, 61
thermodynamics, 97
TOE, vii
total mass, 28, 29, 75
transparent phase of the expansion,
    80
transparent phase of the universe, 56
tunneling from nothing to something,
    12
two-dimensional space, 38
Type Ia supernovae, 65

uncertainty, 90, 91
undeniable differences
    alphabet, 97
    appreciate the beauty, 97
    great buildings, 97
    languages, 97
    paintings, 97
    symphonies, 97
universal, 94
universal constants
    Boltzman, 17
    Newton, 17
    Planck, 17
    universe, 17
    velocity of light, 17
universe, 5–7, 11, 18, 19, 22, 24, 25,
    27–29, 31, 32, 38, 39, 41, 42, 45–47,
    49–51, 55–57, 59–61, 65, 71, 72, 75,
    77–81, 85–87, 89–91, 93, 98, 102,
    121–125
    intelligible, 102
    matter energy content, 73
Uranus, 94

velocity versus redshift, 66
Venus, 94, 95

water, 10, 94, 95
white dwarf, 125, 126
WIMPS, 72
WMAP, 75, 98, 102
World War One, 59, 101
World War Two, 87, 102

X-rays, 21, 22

yin and *yang*, 90

zero-point energy of the vacuum, 73